MW01641179

PREVENTING NUCLEAR MELTDOWN

Preventing Nuclear Meltdown

Managing Decentralization of Russia's Nuclear Complex

Edited by

JAMES CLAY MOLTZ
VLADIMIR A. ORLOV
ADAM N. STULBERG

ASHGATE

Published by
Ashgate Publishing Limited
Gower House
Croft Road
Aldershot
Hants GU11 3HR
England

Ashgate Publishing Company
Suite 420
101 Cherry Street
Burlington, VT 05401-4405
USA

Ashgate website: http://www.ashgate.com

British Library Cataloguing in Publication Data
Preventing nuclear meltdown : managing decentralization of
Russia's nuclear complex
1. Nuclear nonproliferation 2. Nuclear facilities - Russia
(Federation) - Management 3. Nuclear facilities - Russia
(Federation) - Management - Case studies 4. Nuclear
facilities - Security measures - Russia (Federation)
5. Nuclear facilities - Security measures - Russia
(Federation) - Case studies 6. Regionalism - Russia
(Federation) - Case studies
I. Moltz, James Clay II. Orlov, Vladimir A. III. Stulberg,
Adam N.
327.1'747'0947

Library of Congress Cataloging-in-Publication Data
Preventing nuclear meltdown : managing decentralization of Russia's nuclear complex /
James Clay Moltz, Vladimir A. Orlov, Adam N. Stulberg, editors.
p. cm.
Includes index.
ISBN 0 7546 4257 7
1. Nuclear weapons industry--Russia (Federation) 2. Nuclear power
plants--Location--Russia (Federation) I. Moltz, James Clay. II. Orlov, Vladimir A. III.
Stulberg, Adam N., 1963-

HD9744.N833R93 2004
338.4'762145119'0947--dc22

2004046270

ISBN 0 7546 4257 7

Printed and bound in Great Britain by MPG Books Ltd, Bodmin, Cornwall

Contents

PART III: THE EXPERIENCE OF U.S. ASSISTANCE PROGRAMS AND CONCLUSIONS

List of Figures and Tables

List of Contributors

Sonia Ben Ouagrham

Dr. Sonia Ben Ouagrham is an economist specializing in the conversion of the defense industry in the former Soviet Union (FSU). She received her Ph.D. in Economics of Development from the Advanced School of Social Sciences in Paris, France. Currently, she is a senior project manager at the Center for Nonproliferation Studies' Washington office, and editor-in-chief of the *NIS Export Control Observer*, a monthly newsletter devoted to the analysis of WMD export control issues in the NIS. From 1999-2001, Dr. Ben Ouagrham was research coordinator at the center's office in Almaty, Kazakhstan. During that period, she conducted a study on the restructuring of the nuclear sector in Russia. Dr. Ben Ouagrham was a key contributor to *Anatomy of the Russian Conversion* (2001, edited by Vlad Genin). She has also conducted extensive research on issues related to conversion of biological weapons facilities in Russia, Kazakhstan, and Uzbekistan.

Cristina Chuen

Cristina Chuen is a senior research associate at the Center for Nonproliferation Studies of the Monterey Institute of International Studies, where she also serves as coordinator of the NIS Nuclear Profiles Database. She is an expert on issues related to Russia's civilian and naval nuclear reactors, foreign assistance developments, and activities under the Russian Ministry of Atomic Energy. She is a Ph.D. candidate in International Affairs at the University of California at San Diego and holds an M.A. in Russian and Chinese history at the University of Hawaii and an A.B. in Soviet Studies from Harvard University. She is the author of recent articles in such publications as the *Bulletin of the Atomic Scientists*, the *Los Angeles Times*, and the *South China Morning Post*.

Dmitry Evstafiev

Dr. Dmitry Evstafiev is a senior research associate at the PIR Center, where he is also editor of the journal *Yaderny Kontrol* and director of the program on "Russian Domestic Politics & Security." He is a graduate in History from Moscow State University and holds a doctorate in Political Science from the USA and Canada Studies Institute of the Russian Academy of Sciences. From 1992-95, Dr. Evstafiev was an analyst at the USA and Canada Studies Institute and from 1995-98 was a senior fellow at Russia's Institute for Strategic Studies. He serves as an editor of the journal *New Russia* and is chief of the editorial board of the journal

Deti Feldmarshal (Field-Marshal's Children). He is the co-author of two books and has written many articles for publication in newspapers and scientific and political journals within Russia.

Michael Jasinski

Michael Jasinski is a former arms control inspector for the U.S. On-Site Inspection Agency (currently part of the Defense Threat Reduction Agency) for the Threshold Test Ban Treaty and the Conventional Forces in Europe Treaty. His experience also includes implementing various Cooperative Threat Reduction Program projects in the Russian Federation, Ukraine, and Belarus. He has an M.A. in Russian and East European Studies from the University of Kansas and is currently pursuing a Ph.D. in Political Science at the University of Georgia. Until December 2003, he was a research associate at the Center for Nonproliferation Studies of the Monterey Institute of International Studies.

Dmitry Kovchegin

Dmitry Kovchegin is a graduate of the Moscow Engineering and Physics Institute (MEPhI) and is currently a Ph.D. candidate at the Institute of World Economy and International Relations (IMEMO) of the Russian Academy of Sciences. He works as a research consultant to the U.S.-Russian Nonproliferation Working Group in Moscow. Kovchegin has held past fellowships at the Belfer Center for Science and International Affairs at Harvard University and the Center for Nonproliferation Studies of the Monterey Institute of International Studies.

James Clay Moltz

Dr. James Clay Moltz is deputy director and research professor at the Center for Nonproliferation Studies (CNS) of the Monterey Institute of International Studies. Until August 2003, he directed the Newly Independent States Nonproliferation Program at CNS. From 1993-98 he was the Founding Editor of *The Nonproliferation Review* and from 1996-2000 also of the web-based *DPRK Report*. Dr. Moltz is co-author of the book *Nuclear Weapons and Nonproliferation* (2002), co-editor of *The North Korean Nuclear Program: Security, Strategy and New Perspectives from Russia* (2000), and author of articles in such journals as *Arms Control Today*, *Asian Survey*, *Bulletin of the Atomic Scientists*, *Journal of East Asian Studies*, *NBR Analysis*, and *World Politics*. He holds a Ph.D. in Political Science from the University of California at Berkeley and an M.A. in Russian and East European Studies and B.A. in International Relations from Stanford University. He worked previously as a staff member in the U.S. Senate and has served as a consultant to the U.S. Department of Energy.

Vladimir A. Orlov

Dr. Vladimir A. Orlov is the founding Director of the Moscow-based Center for Policy Studies (PIR Center), where he also directs the "Nuclear Nonproliferation & Russia" Program, and serves as editor-in-chief of the center's journal *Yaderny Kontrol* (Nuclear Control). Since January 2004, Dr. Orlov has been a faculty member at the Geneva Center for Security Policy (GCSP). Dr. Orlov was the vice president, a member of the Board of Directors, columnist, political analyst, and head of department for *Moskovskiye Novosti* (Moscow News) from 1990-1996. He is a Ph.D. graduate of the Moscow State Institute for International Relations (MGIMO). He has written articles for the *Bulletin of the Atomic Scientists*, *Nezavisimoye Voyennoye Obozreniye* (Independent Military Review), *Vremya MN* daily, *Itogi* magazine, *The Nonproliferation Review*, *Krasnaya Zvezda* daily, *Pro et Contra* quarterly, and other Russian and international publications. He has also edited several books and written chapters for several books on nonproliferation published in the West and in Russia. Dr. Orlov has served as a consultant to the United Nations on nonproliferation education.

Ivan Safranchuk

Ivan Safranchuk is director of the Moscow Office of the Center for Defense Information (CDI). Before joining CDI, he worked from 1997-2001 as a research fellow and then "Nuclear Arms Control" project director at the Center for Policy Studies in Russia (PIR Center). He has published numerous articles in Russian and English on nuclear arms control, WMD nonproliferation, and Russian foreign policy. He is a graduate of the Moscow State Institute of International Relations (MGIMO).

Nikolai Sokov

Nikolai Sokov is a senior research associate at the Center for Nonproliferation Studies of the Monterey Institute of International Studies. He graduated from Moscow State University in 1981 and subsequently worked at the Institute of USA and Canadian Studies and the Institute of World Economy and International Relations in Moscow. From 1987-92 he worked in the Ministry of Foreign Affairs of the Soviet Union and Russia, dealing with nuclear arms control; where he participated in START I and START II negotiations, as well as in a number of summit and ministerial meetings. Dr. Sokov holds Ph.D. degrees from the University of Michigan and the Institute of World Economy and International Relations (IMEMO). Sokov has published extensively on international security and arms control and is the author of the book *Russian Strategic Modernization: Past and Future* (2000).

Elena Sokova

Elena Sokova is director of the Newly Independent States Nonproliferation Program at the Center for Nonproliferation Studies of the Monterey Institute of International Studies. Before coming to the United States, Elena Sokova worked for the Russian Ministry of Foreign Affairs. She graduated from the Moscow State University Law School and also holds an M.A. in Public Administration from the Monterey Institute of International Studies. Her research focuses on nuclear nonproliferation issues in Russia's regions, fissile material production and disposition in Russia, nuclear smuggling in the NIS, and foreign nonproliferation assistance to Russia and the NIS.

Adam N. Stulberg

Adam N. Stulberg is an assistant professor of International Affairs at the Sam Nunn School of International Affairs at Georgia Institute of Technology. He specializes in Russian and Eurasian foreign and security policies, international security, energy security, and the politics of WMD proliferation. Dr. Stulberg earned his Ph.D. in Political Science from the University of California at Los Angeles. He also holds an M.A. in International Affairs from Columbia University and a B.A. in History from the University of Michigan. He has served as a consultant to RAND, former Senator Sam Nunn, and to the Department of Defense's Office of Net Assessment.

Charles Thornton

Charles Thornton is a doctoral candidate in the School of Public Affairs at the University of Maryland, where he also serves as a graduate research fellow in the Center for International and Security Studies, focusing on arms control and nonproliferation issues. As an employee of Science Applications International Corporation (SAIC), Thornton spent over six years helping the U.S. Department of Defense (DOD) manage the Nunn-Lugar Cooperative Threat Reduction Program. He continues to consult for DOD through SAIC. He received an M.A. from The George Washington University in Security Policy and a B.A. in Political Science from the University of California at Berkeley.

Key Nuclear-Related Sites in Russia

Chapter 1

Introduction: Russia's Nuclear Regions

James Clay Moltz[1]

Managing the nuclear complex across the vast expanse of the Soviet Union posed difficult challenges to the Soviet leadership. But by the combined use of several intersecting measures—strict controls over the movement of people, abundant physical protection in the form of fences and armed guards, and the ability of political police to arrest facility personnel and potential criminal perpetrators on the slightest suspicion—the Soviet Union effectively kept its nuclear energy and defense-related assets safe from would-be smugglers and terrorists. With the break-up of the Soviet Union, however, borders opened, many guards lost their wages and disappeared, fences fell into disrepair, and central funding for scientists collapsed. At the same time, an encouraging but unruly effort at democratization was transferring powers wholesale away from the center during the years of President Boris Yeltsin, despite gaps in the abilities of new, sub-national bodies to handle their new responsibilities. In the nuclear sector, these conditions led to serious problems by the early to mid-1990s, when financial reserves left over within the system from the Soviet era had been used up, critical staff had drifted away, and inappropriate regional actors (including criminal groups) had begun to step into the yawning gap in regional authority.

Today, despite the centralizing reforms introduced beginning in 2000 by Russian President Vladimir Putin, problems remain in managing nuclear facilities located in Russia's distant regions. The primary reason is financial, since these facilities have been much harder hit by Russia's economic decline than central facilities in and around Moscow. Where resources are lacking from the federal level, regional entities (governmental, commercial, or social) often have to pick up the slack or simply make do with lower levels of funding. This poses particular problems in the nuclear sector, given the risks insecure facilities represent to

[1] The author thanks participants in a seminar organized by Bruce Parrott at the Johns Hopkins School of Advanced International Studies in Washington, D.C., for their useful comments and suggestions on an initial draft of this chapter. The author also expresses his appreciation to Caitlin Baczuk for her expert editorial assistance in the final production of this manuscript.

Russian and global security. A second reason is the still-struggling evolution of Russian democracy, where considerable decisionmaking power has devolved to Russia's regions, but effective institutions to manage these new responsibilities do not yet exist. Governmental expertise in the nuclear field is often absent, industry experts are reluctant to admit to problems within their facilities, non-governmental organizations (NGOs) are few and vulnerable to state pressure, and the media is often weak, poorly funded, and under the control of local enterprises, oligarchs, or political machines. These conditions make rational decisionmaking and information gathering on nuclear matters virtually impossible, promoting instead lowest-common-denominator decisions made largely on the basis of local needs—sometimes to the detriment of reactor safety, the security of nuclear materials, and national welfare.

These issues are of particular concern given the known interest of international terrorists in the vast stockpile of nuclear weapons, separated weapons-grade material, and radioactive spent fuel and waste (suitable for so-called "dirty bombs") spread among many dozens of sites in the still-sprawling Russian nuclear complex. In this context, the failure of existing foreign assistance programs, including in the United States, to craft focused strategies for dealing with regional problems represents a weak link in the war on terrorism and efforts to bring about meaningful conversion and downsizing of the Russian nuclear complex. Regional factors can no longer be viewed as "external" to foreign assistance programs. Instead, in the evolving Russian political context, they have to be understood as having significant and direct effects on their success.

The fact that regionally based problems in the nuclear sector have not been dealt with in a systematic way by Western policymakers or academics is not entirely surprising. Indeed, there were understandable reasons for this neglect. Part of the problem was a lack of information regarding conditions at difficult-to-access regional facilities. But part was also the intellectual challenge of coming to grips with the unprecedented concept of possible "regionalism" in regards to nuclear energy and weapons-related material, an oxymoron in the context of strictly hierarchical Soviet political and economic structures.

Tearing down the old Soviet state was a long-sought goal of many Western governments—as well as the hope of many scholars studying the Soviet Union. But the prospect of nuclear authority devolving to a dizzying number of emerging regional authorities as a "cost" of de-Sovietization was too much for many to fathom after decades of rigid central control. In the end, fortunately, the emerging reality of waning central authority and growing new regional influence in the nuclear sector (often born of necessity, rather than any evil intentions) has developed somewhere in between the two extremes. Most regions have not been eager to seize complete nuclear authority, but most also are still not able to handle even those limited functions for which they have become responsible.

Problems were probably inevitable, given the conditions that emerged in Russia's regions in the early 1990s. As two Norwegian authors sum up the net effects of economic and political changes on the Russian nuclear complex as witnessed in facilities located on the Kola Peninsula: "The flow of resources from the federal center to the region decreased, and the poverty of the military complex increasingly became a problem which Murmansk oblast itself had to deal with."[2] Unfortunately, as several cases of nuclear theft highlighted, Murmansk (and many other regions) did not prove up to the task. In many areas, such capacities continue to be inadequate, in some cases, dangerously so.

Nevertheless, particularly since Putin's rise, there is a widespread tendency among Western governments to assume that firm centralized control over nuclear decisionmaking in Russia is still more or less in place. Such assumptions, however, are at best exaggerated and, at worst, mask a misunderstanding of current dynamics within the Russian nuclear complex. Although conditions tend to vary between weapons and non-weapons facilities (with the former usually being safer), there is abundant evidence that previously reliable central control mechanisms in Russia's nuclear complex have come undone, and regional solutions have been far less than optimal. City governments have taken over critical supply functions for the Russian nuclear navy, governors are ordering that power plants run beyond their suggested service lives due to desperate local needs for heat and electricity, and weakened safety regimes in particular areas are bringing drugs and mafia groups into closed nuclear cities as never before. These problems pose threats not only in terms of the safety of local populations, which suffer from inadequate protection *from* dangerous nuclear materials,[3] but also proliferation concerns.

In the context of the ongoing U.S. and European struggle against terrorism, the dangers these conditions represent are clear: a continuing risk of the proliferation of fissile material and weapons-related technologies from regional facilities, a growing possibility of catastrophic nuclear accidents due to uninformed local decisionmaking, and the failure of Western nonproliferation assistance programs to achieve their goals due to misplaced trust in federal authorities or an inability to predict local influences. A related issue is the emerging alliance between Minatom and regional governors built around their common desire to expand nuclear power production, foreign spent fuel storage, and nuclear reprocessing. These trends could result in increased plutonium production in Russia and facilitate development of a plutonium-based fuel cycle, generating even more serious long-term problems by boosting the stock of weapons-grade material within an already at-risk system. While certain aspects of the growth of regional influences do indicate the emergence of greater democratization in Russia, the nature and

[2] Geir Honneland and Anne-Kristin Jorgensen, *Integration vs. Autonomy: Civil-Military Relations on the Kola Peninsula* (Aldershot, UK: Ashgate Publishing, 1999), p. 126.

[3] One example is the radioactive material that powers many lighthouses in Russia. In some cases, it is no longer protected by guards, and thieves have seized these source materials in an attempt to sell them for profit. In several cases, the thieves themselves have been hospitalized. In others, children who have been playing at unguarded lighthouses have received high doses of radiation and have required extensive medical treatment.

direction of decisionmaking may not always be to the liking of outside assistance providers seeking to reduce Russia's nuclear infrastructure and exports. For these reasons, understanding the ramifications of Russia's nuclear decentralization and/or "regionalism" requires further research and attention by scholars and policymakers alike.

This book analyzes the significant changes in the governance and operation of Russia's regional nuclear facilities since 1991, which have been quite remarkable in the context of Soviet-era practices. But most of these developments have been organic rather than planned, leading to conditions ripe for negative manifestations of regionalism. More generally, we are witnessing the widespread emergence of decentralized and sub-national influences on the operation of nuclear facilities, particularly in the formulation *and* implementation of specific policies. Thus, the concepts of decentralization and regionalism used in this volume broadly encompass the emergence of a wide range of elected, appointed, and private actors below the federal level, as well as the social and economic factors that together have caused an unprecedented devolution of decisionmaking authority within the Russian nuclear complex. Not all of these developments are necessarily negative, as certain sub-federal level actors have on occasion intervened to help stabilize the situation at specific facilities, filling commercial and administrative voids created by the uncertainty in federal oversight and governing practices. Moreover, the evidence from various regions is not always consistent and regional peculiarities abound. For these reasons, a comprehensive study of the different faces of emerging regional influences in Russia's nuclear sector is needed to categorize, analyze, and take stock of how Soviet power has been replaced by a complex array of new factors.

This book represents the culmination of a study initiated in the spring of 2000 by a team of U.S. and Russian scholars led by the Monterey Institute of International Studies' Center for Nonproliferation Studies and the Moscow-based Center for Policy Studies in Russia (PIR Center), an independent NGO. Besides analyzing emerging center-periphery relations in the nuclear sector, the project team studied the behavior of the Ministry of Atomic Energy and the Russian military in the regions, focusing in particular on five critical areas with large nuclear infrastructures inherited from the Soviet period: the Northwest (including St. Petersburg and Murmansk), the Volga region (including Nizhniy Novgorod and Sarov), the Urals (including Snezhinsk and Mayak), Siberia (including Zheleznogorsk and Seversk), and the Far East (including the Petropavlovsk and Vladivostok areas).

Data from these case studies showed a number of troubling findings, including extensive theft from regional nuclear facilities, the breakdown of safety procedures, pro-nuclear power sentiments among regional governors, extensive mafia activity in and around nuclear facilities, and evidence of foreign intelligence activities from countries of concern at facilities in border regions. But they also showed that—in some areas—regional environmental groups were beginning to exert a positive influence, regional governors were cooperating to develop broad-area plans for electricity development, and Moscow-based officials were beginning to recognize the need to take local opinions into account before initiating nuclear expansion (or contraction) efforts. This new and variegated reality suggests that assumptions about the continued exercise of "top-down" central control over regional nuclear facilities was seriously flawed, requiring new attention to regional political and economic factors.

The rest of this introductory chapter briefly reviews the literature on Russian regionalism, noting the absence among these studies—until very recently—of specific attention to critical nuclear security issues. It then examines current U.S. and Western nuclear assistance policies in Russia and describes how structural biases in both Western capitals and Moscow have prevented more concentrated action to address regional problems. Next, the chapter lays out the book's overall structure and the individual contributions. Finally, my conclusion highlights some of the common themes that emerge from the considerable range of evidence presented in this volume.

Fortunately, the data provided by the various authors do not offer clear-cut cases of missing nuclear weapons or the seizure of regional control over any nuclear stockpiles. But they do indicate serious problems in a number of regional facilities with regard to nuclear fuel disposition, waste storage, and safety measures caused by regional dynamics. Many of these difficulties relate to the fact that Russia's weakened, post-Soviet center—even under Putin—must also cope with "even *weaker* regions," rather than the stronger and more capable ones that democratization literature might predict.[4] Thus, while regional influence over nuclear decisionmaking has increased, the availability of resources to improve conditions at local nuclear facilities is less than during the Soviet period.

Existing Studies of Russian Regions

In order to understand the forces that have resulted in growing regional influence in Russia's nuclear sector today, it is instructive to step back and review Jerry Hough's seminal study from the late 1960s, *The Soviet Prefects: The Local Party Organs in Industrial Decision-Making*.[5] Using Hough's work, it is possible to set at least a rough baseline of regional power during the Soviet era in order to observe the range of changes that have occurred since 1991.

---

[4] On this point, see Natan M. Shklyar, "Russian Regions and Russia's Foreign Policy," *The Soviet and Post-Soviet Review* 25, No. 3 (1998), p. 286; Shklyar also cites a 1998 study by Thomas E. Graham that makes a similar argument.

[5] Jerry F. Hough, *The Soviet Prefects: The Local Party Organs in Industrial Decision-Making* (Cambridge, MA: Harvard University Press, 1969).

Hough outlined in impressive detail (particularly considering the severe limits of information at the time) the amount of leeway granted to regional industrial officials and Communist Party leaders during the Soviet period in handling key management decisions for Russian regional enterprises. On the plus side, he pointed to the importance of personal contacts within the system and the principle of *edinonachaliye*, or the notion that local leaders were empowered to make crucial management decisions about daily enterprise life, particularly if such decisions contributed to central Communist Party directives, such as the fulfillment of the yearly economic plans. However, on the minus side, Hough's study also showed that on issues of critical importance (whether for political, economic, or security reasons), senior party officials could "reach down" and order compliance by regional "prefects" with the full expectation that they would obey those orders. Such was the fear inherent in the Communist system and so great were the means of retribution available to senior officials against independent thinkers or recalcitrants at each level of the hierarchy that no one dared "crossing" his superiors once an edict had been issued from above. Thus, any independence—even in civilian industries—was purely technical and did not extend to questions of overall policy. In the military area, there was even less room for deviation or maneuver.

At the same time, facilities serving the military sector demanded (and received) extraordinary financial resources under the Communist system in return for their loyalty, which allowed leaders at the local level to succeed in their tasking and to maintain conditions of safety (at least in terms of reliable physical protection over material). Thus, facilities in the nuclear complex benefited disproportionally compared to other industrial sectors, and regions with large numbers of such facilities obtained extraordinary regional benefits in the form of access to hard-to-get industrial inputs, foodstuffs, salary incentives, vacations, and other amenities. In these respects, conditions in the Soviet-era nuclear sector more closely mirrored those in the U.S. nuclear sector than other areas of the Soviet economy, where lack of investment, frequent shortages, rampant corruption, and other workplace problems were allowed to fester, keeping living standards low. Eventually, it was central neglect of these issues—combined with ineffective and heavy-handed management techniques—that led to the Soviet Union's economic decline and contributed to its eventual political break-up.

It is useful to reread Hough's study in the light of today's changes because its findings highlight the tremendous distance that Russia has traveled since the height of Soviet power, and particularly in the decade since 1991. With the collapse of the old system, it can be argued that emerging regional factors had a greater effect on the nuclear infrastructure than they did on other parts of the economy, precisely because this complex had been so protected and therefore had that much further to fall. Unfortunately, the consequences are also much more serious for national and international security.

Political-Economic Studies of Russian Regionalism

In the early 1990s, analysts of the regional role in Russia's political and economic reforms rose to the forefront of Western academic studies, focusing attention on the dramatic breakup of the Communist Party structure, the demise of the state's central planning apparatus for the economy (Gosplan), and the replacement of central ministries with privatized local enterprises. Yet, issues in the nuclear complex remained under-studied. Part of the reason lay in the fact that most of the researchers conducting this work were mainstream political scientists or economists, who naturally focused on issues more central to their disciplines. A number of young scholars conducted pioneering research on various regions and the complex coalitions and local bargaining strategies many developed for dealing with the center in conditions of resource scarcity.[6] Yet, as noted above, their findings did not always show that regions were becoming stronger.[7]

Specialists on Russia's nuclear complex, by contrast, tended to remain focused on the center, where nuclear weapons and arms control decisionmaking continued to be made. Throughout the 1990s, however, authors noted the decline of nuclear safety and Russia's growing inability to handle its Soviet nuclear inheritance (much less pay for arms control reductions). Yet, the regional dimension was not discussed in any organized manner. One exception to this rule was a 1994 article by Jessica Stern, who made the provocative claim that Russia would soon break up into mini-fiefdoms with nuclear arsenals, citing the growth of regional independence movements and the success of the Eastern bloc countries in breaking with Moscow.[8] Although Stern rightly highlighted emerging regional problems, Russia did not end up following her predicted path of disintegration, as dynamics within the Russian Federation proved different from those in other former union republics. Over the course of the 1990s, instead, Russia's regions turned out to be more interested in reacquiring dwindling economic benefits from the center than in achieving total political (and with it nuclear) independence.[9] As a result, most analysts lost interest in regional nuclear questions and turned their attentions elsewhere.

---

[6] See, for example, Daniel S. Treisman, *After the Deluge: Regional Crises and Political Consolidation in Russia* (Ann Arbor, Mich.: The University of Michigan Press, 1999); Mikhail A. Alekseev, ed., *Center-Periphery Conflict in Post-Soviet Russia: A Federation Imperiled* (New York: St. Martin's Press, 1999); Peter Kirkow, *Russia's Provinces: Authoritarian Transformation versus Local Autonomy?* (New York: St. Martin's Press, 1998); and Kathryn Stoner-Weiss, *Local Heroes: The Political Economy of Russian Regional Governance* (Princeton, N.J.: Princeton University Press, 1997).

[7] Shklyar, "Russian Regions and Russia's Foreign Policy."

[8] Jessica Eve Stern, "Moscow Meltdown: Can Russia Survive?" *International Security* 18 (Spring 1994).

[9] The one major exception to this rule, of course, was Chechnya. Tuva and Tatarstan also showed some pro-independence tendencies, but eventually proved willing to work with Moscow within the federation.

In the meantime, considerable useful research on new Western (mainly U.S.) nonproliferation assistance programs and the dangerous nuclear legacies in newly independent Belarus, Kazakhstan, and Ukraine began to fill policy journals and provide a wealth of information on new proliferation threats in the former Soviet Union.[10] Many of the nuclear analysts writing these articles recognized the emergence of control problems at Russian facilities as well. But most weapons analysts lacked the detailed knowledge of Russia's emerging regional politics and economics required to study emerging nuclear factors below the national level. Moreover, existing biases in the field based on assumptions of enduring central control in this key area of Russian security—supported by exaggerated statements from Moscow-based officials (themselves often ignorant of changing conditions in the regions)—diverted attention from emerging center-periphery problems in the nuclear sector. Finally, from a purely practical perspective, uncovering these developments required travel to distant and hard-to-reach locations, the collection of sensitive data on facilities, and asking uncomfortable questions of regional officials about nuclear materials. Such obstacles blocked new research, at least at first.

During the late 1990s, however, analysts with an interest in civil-military relations began to conduct some related work. Noticing a number of anomalous patterns at regional facilities—compared to the Soviet period—these scholars described the emergence of new relationships between military and regional elites. As Dale Herspring observed, contacts had emerged in many locations between the military and local economic and political leaders that "...have serious implications for the nature of civil-military relations in the country."[11] Eva Buzsa identified and described a new and increasingly widespread phenomenon in which regional governments were beginning to pick up social costs for military units, including such necessities as housing and food. She concluded, "These developments are important because there is every reason to expect that economic ties will translate into political loyalties."[12] Still, no focused attention was paid to the vast yet hard-to-penetrate nuclear complex itself.

---

[10] See, for example, the special issue of *Arms Control Today* on "Nuclear Weapons In the Former Soviet Union," January/February 1992; also the volume *Cooperative Denuclearization: From Pledges to Deeds*, Allison, Carter, Miller, and Zelikow, eds., CSIA Studies in International Security No. 2 (January 1993); Mark D. Skootsky, "An Annotated Chronology of Post-Soviet Nuclear Disarmament 1991-1994," *Nonproliferation Review* 2 (Spring-Summer 1995); and the book *Dismantling the Cold War*, Shields and Potter, eds. (Cambridge, MA: MIT Press, 1997). This is a just a small sampling of the many useful studies published on these topics.

[11] Dale R. Herspring, "The Russian Military Faces 'Creeping Disintegration'," *Demokratizatsiya* 7 (Fall 1999), p. 581.

[12] Eva Buzsa, "From Decline to Disintegration: The Russian Military Meets the Millennium," *Demokratizatsiya* 7 (Fall 1999), p. 563.

In early 1999, a few private contractors with extended experience in implementing U.S. nonproliferation assistance programs in Russia—who had noticed the emergence of new regional obstacles while trying to carry out their work—began discussing these questions with analysts at the Center for Nonproliferation Studies (CNS) at the Monterey Institute. Separately, former Soviet Ambassador to the International Atomic Energy Agency Roland Timerbaev of the PIR Center also began to notice problems in Russia's regions in regards to nuclear security. Intrigued and troubled by some of their reports, CNS and PIR Center staff teamed up with several independent scholars working on related subjects to begin investigating these issues. Encouraged by CNS Director William Potter and inspired by some useful suggestions from Nikolai Sokov, the present author undertook a pilot study for this project in an area where he had previously conducted considerable research: the Russian Far East. His findings from reading the local press and conducting interviews with experts and officials in Vladivostok and Khabarovsk in the fall of 1999 proved surprising and troublesome. He discovered considerable evidence of emerging regional influence in the nuclear sector, including unprecedented strikes by workers, contract killings at key nuclear installations, and failure of local guard forces to keep out foreign and mafia influences at nuclear sites. The author discussed these problems in conference presentations and later published an initial study on nuclear regionalism in the Far East.[13] Drawing on the nuclear weapons literature, he categorized three types of possible regional influence over nuclear facilities: 1) "positive" control (seizure of weapons, material, or even whole enterprises); 2) "negative" control (actions that prevented the center from making use of regional nuclear assets); and 3) "loss of control" situations (where control had broken down so much that the center no longer could exercise control over regional nuclear facilities).[14] Within the Far East, the author was able to offer data that matched each of these three types of nuclear regionalism. At the same time, other analysts unconnected with this emerging project also began to investigate related regional security themes. One result was a book about developments on the Kola Peninsula by a team of Norwegian analysts, which covered a number of important nuclear issues.[15]

Then, in early 2000, an article published in *Foreign Affairs* by former Senator Sam Nunn and Adam Stulberg of the Georgia Institute of Technology[16] raised a similar set of concerns in the broader context of U.S. security. They described an ongoing center-periphery "tug of war" in the Russian nuclear sector that was likely to continue for some time. Given this ongoing "drift" in relations between center

---

[13] James Clay Moltz, "Russian Nuclear Regionalism: Emerging Local Influences over Far Eastern Facilities," *NBR Analysis* 11 (December 2000).

[14] Ibid.

[15] Honneland and Jorgensen, *Integration vs. Autonomy: Civil-Military Relations on the Kola Peninsula*.

[16] Sam Nunn and Adam Stulberg, "The Many Faces of Modern Russia," *Foreign Affairs* 79 (March-April 2000).

and periphery in Russia, they argued, it was impossible for Western assistance programs working primarily through central organizations to be effective without taking these influences into account. As Nunn and Stulberg noted, "Classic notions of realpolitik and nation-state interplay are growing less and less relevant." They added, "Given this shifting balance of power from the center to the periphery, policy-makers in Washington must rethink America's strategy for engaging Russia...."[17]

While President Vladimir Putin tried to rein in the rebellious regions in his spring 2000 reforms, he failed to provide the kind of financial or political support needed to make his programs effective and to cement central authority. The appointment of the seven presidential super-representatives to the regions had reduced the number of presidential envoys overall, thus presumably making each one more powerful. The removal of governors from the Federation Council had also reduced regional levers over central legislation and boosted the relative power of the president and the State Duma. Certain budget reforms also guaranteed greater tax receipts for the center. But the impact of the reforms did not correlate directly with greater central control over regional governors, who began to co-opt the envoys to serve as their lobbyists in the capitol. Similarly, Putin's reforms did little to replace emerging regional influence in the economic sector, nor did they reestablish the old vice-grip of central ministries over local industrial enterprises. Ironically, by clamping down on regional tax receipts and other financial instruments, the reforms seemed only to *exacerbate* problems of security in the regional nuclear enterprises.

With funding from the Smith Richardson Foundation, a joint CNS-PIR Center project to conduct an extensive study of nuclear regionalism across Russia began in the summer of 2000. Eventually, more than a dozen researchers conducted studies in six major regions of Russia, including hundreds of interviews with regional and national officials, as well as analysis of the activities of federal-level bodies in the regions (including Minatom and the military). Extensive reading of regional press sources also played an essential role in this study. Each team member was asked to examine a common set of questions as applied to his/her region or topic, including: the nature of political-economic relations with the center; the relative weight of organized crime and its relationship to regional actors and nuclear facilities; the region's experience with foreign nonproliferation assistance; the stability of key inputs to the nuclear sector (personnel, electricity, water, and heat); the role (if any) of NGOs; and specific incidents that highlight problems of nuclear safety and control.

Despite its scholarly intentions, the project also aimed at producing tangible results for policymakers. Indeed, three meetings were held (two in Washington and one in Moscow) to brief government officials, experts, and media representatives

---

[17] Ibid., pp. 46–47.

on findings from the project. In addition, authors endeavored to provide policy recommendations as part of their conclusions.

Western and Russian Policy Responses to Russia's Regional Problems

When Representative Les Aspin and Senators Sam Nunn and Richard Lugar came up with the idea in fall 1991 of providing assistance for the purpose of promoting arms control and nonproliferation to the then-disintegrating Soviet Union, it was a radical concept. During the course of the 1990s, however, these policies became widely accepted as critical to U.S. security, leading eventually to the U.S. investment of over $3.5 billion during the decade for the so-called "Nunn-Lugar" program (including activities in the Departments of Defense, State, Commerce, and other agencies).

Since 1991, U.S. assistance programs in the former Soviet Union have aimed at: 1) facilitating the elimination of weapons of mass destruction (WMD); 2) reducing chances for the proliferation of WMD know-how and material (including through training programs and providing non-weapons research work to weapons scientists); and 3) assisting in the safeguarding and storage of nuclear materials to facilitate goals number 1 and 2. Due to the nature of the government-to-government agreements covering these programs, U.S. policy had tended to work through the Moscow-based ministries and the central command staff of the military to achieve its goals. By the late 1990s, these programs had done a great deal to alleviate many current threats. But they had left other problems unaddressed and began to bump into increasingly troublesome issues stemming from regional influences.

Due to bureaucratic pressures, the difficulties of working in Russia, and the time-consuming process of developing trusted personal ties with central authorities, however, U.S. government officials showed great reluctance to reach beyond their existing Moscow-based or laboratory-based networks and to engage Russian local and regional officials. These trends persisted despite the fact that many of them had struggled through difficulties involving illegal or "supplemental" taxation at the regional level, denials of site access, illegal diversions of materials, sales of technologies without licenses, and related concerns that had put their programs at risk. Indeed, while many officials had run into obstacles clearly rooted in regional political and economic problems, there was no system-wide approach within the U.S. government for dealing with these difficulties and no mechanism for gaining information to prevent them in the future. Instead, in many cases, institutional factors caused U.S. officials to obscure or downplay information from the regions that cast doubt on central Russian players' control over regional facilities, so as not to jeopardize Russian government or U.S. Congressional support. But the reasons for the failure to address these problems had deeper, structural roots that shed some light on the nature of the assistance programs themselves—and are therefore worth discussing.

To begin with, the vast majority of U.S. government officials working in nuclear assistance programs were technical specialists. While extremely knowledgeable in nuclear issues, they lacked expertise in historical and even geographic information about Russian nuclear facilities. With one or two exceptions (out of dozens), program managers and senior officials did not speak Russian. From extensive interviews at the time, it was clear that most officials lacked basic information on the political and economic realities of the regions where their projects were located.[18] With some exceptions, therefore, U.S. nuclear assistance officials tended to rely on their Russian counterparts in Moscow or within laboratories in the regions as their primary sources of information. While this was somewhat understandable, particularly given the high level of trust that was needed to establish relations with central Russian officials and team partners, it sometimes led to faulty assumptions about political and economic conditions on the ground and the ability of Russian officials to deliver on their promises. Within Russia's Minatom, for example, there was an extreme reluctance to admit to U.S. authorities the ministry's weakening hold over regional facilities, despite the emergence of difficulties. The rising independence of regional governors and enterprises was an unpleasant reality, but a reality nonetheless. Yet, the seriousness of the problem of nuclear regionalism was often kept from political leaders in Russia too, including in the presidential administration. There was also little active discussion of these problems in the Russian Duma, much less in the U.S. government, with a few exceptions.[19]

Frustrated by what they perceived as a failure in the Clinton administration to deal adequately with problems within U.S. nonproliferation assistance programs (many of them in Russia's nuclear regions), U.S. policymakers under the Bush administration initially stepped back in early 1991. In certain respects, this may have been a logical approach for the new administration, wanting to reassess the situation. Yet, initial policies of reducing funds for material protection, control, and accounting programs seemed short-sighted in the absence of proposed alternatives. Former officials like Rose Gottemoeller criticized early cuts in particular

[18] Author's interviews with U.S. officials in the Departments of Energy, Commerce, and Defense, 1995–2001.

[19] Notably, some efforts in this regard in the Department of Defense's Cooperative Threat Reduction (CTR) Program yielded success. Specifically, the 1997 CTR decision to begin direct contracting with Russian shipyards located in the Far North and the Far East allowed funds to flow directly to the facilities responsible for implementing strategic nuclear submarine dismantlement, rather than getting caught up in Moscow. This greatly accelerated the dismantlement rate.

programs, some of which were aimed precisely at trying to alleviate regional problems.[20]

After the attacks of September 11, 2001, in New York and Washington, however, the Bush administration changed its course. It now recognized that fissile material in far-flung Russian facilities under conditions of dubious control could fall into the hands of terrorists, which would create problems that would be far harder (and more costly) to manage later on, such as when a bomb or radiation dispersion device might be deployed against U.S. forces abroad, U.S. friends and allies, or the U.S. homeland.

Today, few voices question the urgent need to ensure the success of these efforts, which are now recognized as playing a crucial anti-terrorist function. But the question is how much attention they deserve relative to other threats? Former senior Reagan administration official Larry Korb argued in February 2004 that the Bush administration was not paying adequate attention to the Russian problem, making the case that "funding for the Nunn-Lugar Cooperative Threat Reduction Program should be increased from its current level of $450 million to at least $2 billion."[21] Yet, the changing face of Russia and the emergence of other global hot spots have injected considerable uncertainty into the calculations of foreign assistance providers, despite the closer relationship of the Putin administration to the United States that has emerged since 9/11.

A considerable time has passed now since the leaders of the Group of Eight (G-8) leading industrialized countries announced at their June 2002 summit in Canada an ambitious plan to provide $20 billion in nonproliferation assistance to Russia and the other former Soviet states aimed at preventing fissile material and other precursors for WMD, related technologies, and former weapons scientists becoming accessible to international terrorists. At the 2003 Evian summit, the parties further elaborated their shared objectives, but no detailed plan emerged regarding how to spend this money or how individual donor activities would be coordinated. Part of the problem is a preference by donors to keep close control over their funds, but part is also related to uncertainty about the ability of the central government in Moscow to follow through on its pledges, ensure cooperation in-country, and manage such large sums of assistance funding responsibly. Within this context, a significant question mark is the role of Russia's sub-national and regional actors—as possible obstructionists or facilitators—on key issues such as access, taxation, power supply, labor provision, and liability. Should assistance providers engage Russia's regions directly or instead rely on the federal government, as is now generally the case?

Today, more U.S. officials in the Departments of Defense, Energy, and Commerce are aware of the regional dimension of Russian nuclear proliferation

[20] See Gottemoeller's quotes in Walter Pincus, "Bush Targets Russia Nuclear Programs for Cuts," *Washington Post*, March 18, 2001.

[21] Larry Korb, "The World As It Isn't" (Op-Ed), *The Wall Street Journal*, February 4, 2004.

problems, but bureaucratic pressures not to "rock the boat" with existing Russian counterparts still frequently result in a tendency in some programs to ignore new information from the regions, particularly when it comes from sources such as NGOs, journalists, or other non-official channels. Unfortunately, it is precisely this kind of regional data (as presented in this book) that is most needed, if U.S. and other Western programs are going to be able to adapt and succeed in the new environment in Russia.

In 2003, Senate Foreign Relations Committee Chairman Richard Lugar supported a redoubling of U.S. efforts to ensure that Russian-origin fissile material and other radioactive sources are safeguarded at their point of origin. Keenly aware of the complexity of the issues from his extensive travels to Russia, he stressed that without addressing the regional dimension of proliferation problems, the United States and other foreign assistance providers would ultimately fail to achieve their objectives. Senator Sam Nunn and his colleagues at the Nuclear Threat Initiative in Washington, D.C., also remained concerned about these issues, and developed specific programs to combat these problems. This was complemented by efforts undertaken at the Princeton University's Center for Energy and Environmental Studies, the Russian American Nuclear Security Advisory Council, the Carnegie Endowment for International Peace, the University of Georgia's Center for International Trade and Security, Harvard University's Managing the Atom Project, and a number of other centers. While also serving the scholarly purpose of providing an objective analysis of the issue of nuclear regionalism in all its dimensions, this book is intended to assist these efforts and improve the effectiveness of U.S. and Western-sponsored assistance programs, as well as Russian policy initiatives, in the field of nuclear nonproliferation.

Structure of the Book and Description of the Chapters

In order to provide as complete a picture as possible of nuclear regionalism in Russia, three sections follow this introduction: Part I (Federal Nuclear Agencies and the Regions), Part II (Case Studies of Russia's Nuclear Regions), and Part III (The Experience of U.S. Assistance Programs and Conclusions).

In Part I, there are four chapters examining various questions related to the role and experience of federal-level political, military, and economic bodies in regards to the Russian nuclear complex and its various regional dimensions. In Chapter 2, Russian security analyst Dmitry Evstafiev and PIR Center Director Vladimir A. Orlov discuss the evolution of center-periphery relations in the Russian political scene and the post-Soviet development of nuclear regionalism. They point to the formal legal and political reasons why nuclear regionalism emerged in the first place, while also providing evidence of some positive trends in recent years. They conclude, however, that we cannot exclude negative manifestations of nuclear regionalism in the future. In Chapter 3, former Soviet and Russian Foreign

Ministry official Nikolai Sokov and Georgia Tech professor Adam N. Stulberg explore the "federal face" of nuclear regionalism in two cases of policymaking: the legislation on spent nuclear fuel imports and proposed changes in the national electricity monopoly. Their work highlights the great activity of regional players at the federal level. Chapter 4, written by economist Sonia Ben Ouagrham (with input from former Russian Foreign Ministry official Igor Khripunov), analyzes the Ministry of Atomic Energy, a critical actor in the nuclear complex and an influential player in Russian politics. They discuss Minatom's many regional relations, but conclude that its long-term strategy—based on domestic nuclear power plant expansion and an ambitious foreign spent fuel storage plan—is not likely to succeed. This section concludes with a study by former On-Site Inspection Agency analyst Michael Jasinski on nuclear weapons and the military in Russia's regions in Chapter 5. Jasinski provides frightening evidence of the desperate state of many military facilities and their increasing need to rely on local and regional (rather than federal) sources of funding. These conditions, he points out, give regional authorities leverage over a variety of activities, including recruitment, training, and even deployments.

In Part II of the book, various U.S. and Russian experts provide detailed case studies of four of Russia's most important regions in regard to nuclear energy, weapons, and fissile material. Chapter 6 traces political and economic conditions affecting nuclear facilities in the Russian Far East, one of the poorest and most isolated regions of Russia and one with some of the most extensive problems. CNS analyst Cristina Chuen's interviews and extensive reading of the local press bring out numerous examples of security breaches at nuclear facilities, particularly those involving the Russian Navy. Chapter 7 by Ivan Safranchuk, head of the Center for Defense Information's branch office in Moscow, discusses regional dynamics in the Volga Federal District. In particular, he tracks the complex politics of Nizhniy Novgorod Oblast and issues surrounding the large All-Russian Scientific Research Institute for Experimental Physics (VNIIEF) in Sarov. After years of neglect by regional authorities, leading to threats of strikes by VNIIEF works, the city seems to be working more closely now with the oblast government in seeking cooperative economic opportunities. However, internal biases within the laboratory and an unwillingness to downsize have created problems for conversion and successful use of foreign assistance. Elena Sokova's analysis in Chapter 8 of the Urals Federal District raises a number of issues of concern. Besides rampant theft and ready access to closed cities such as Snezhinsk (which houses one of Russia's largest nuclear research and design laboratories) to mafia groups, she discusses growing problems of drug use caused by the region's location on major drug smuggling routes from Central Asia. Finally, Russian physicist Dmitry Kovchegin in Chapter 9 covers conditions in the Siberian nuclear cities of Zheleznogorsk and Seversk (each home to a major nuclear fuel cycle facility), examining in particular the national political role of the Siberian Agreement (an association of oblasts and Duma deputies from the region), the prior role of Krasnoyarsk Kray's controversial

Governor Aleksandr Lebed, and the different evolving relationships between these two facilities and their respective regional governments (Krasnoyarsk Kray and Tomsk Oblast).

Part III of the book focuses on U.S. policy concerns and general conclusions. In Chapter 10, Michael Jasinski and former Defense Threat Reduction Agency official Charles Thornton draw on their experiences and extensive interviews with U.S. officials to describe the role of regional factors in the implementation of U.S. nonproliferation assistance programs, focusing primarily on Department of Defense (DOD) and Department of Energy (DOE) programs. Their conclusion is that the different nature of the two programs led to different results. DOD programs have had to deal with more actors and therefore have become more involved in regional politics. Meanwhile, DOE has worked mainly through Minatom and the closed cities, which are, in some cases, more protected from outside influences. Finally, Adam N. Stulberg's concluding chapter examines the multifaceted picture of nuclear regionalism, drawing out evidence, general findings, and policy recommendations from the earlier sections of the book. He argues that aid providers need to recognize the specific characteristics of decentralization in different areas and shape their programs accordingly. In some cases, regional factors may ease problems with federal authorities; in others, they may exacerbate existing disputes between the region and the center—with the result that the aid provider ends up in a dangerous position in between. Such calculations will not be easy, and suggest the need to include many levels of actors before proceeding. However, he cautions that we throw up our hands and ignore these problems at our own peril, given the seriousness and diversity of the nuclear threats some forms of decentralization pose to core Western security interests.

General Themes of the Book

The rest of this book provides detailed evidence and tracks the political and economic context of decentralization and emerging regional influence in Russia's nuclear sector. While highlighting grounds for concern in regards to emerging problems at a number of facilities, as noted above, the authors also emphasize that conditions and problems are not the same throughout Russia. Indeed, some regions have developed unique coping mechanisms (sometimes seen in mutually beneficial relations between oblast and federal authorities) that have allowed them to avoid disasters at specific facilities. Overall, the most serious and widespread problems have to do with safety practices at civilian reactors, the handling and management of spent fuel, and the infiltration of criminal elements into nuclear facilities of many types.

A central finding that runs throughout the chapters—from a political perspective—is that while regional influence may in some cases be linked to greater democracy at the local level, it is not synonymous with it. Moreover, from the perspective of nuclear safety, greater local influence over facilities—in the absence of other developments—is often negative, particularly when the root cause is economic distress (often caused by fiscal abandonment by the center). In these cases, the net result is poorly informed and occasionally dangerous behavior by local decisionmakers that flaunts centrally imposed nuclear safety regulations for the purpose of personal profit, increased energy production, or other rent-seeking purposes. In a few cases, however, there is evidence of greater local control with positive outcomes in terms of oversight of federal decisionmaking and nuclear safety in the context of an increased presence of foreign assistance providers. However, this model is a limited one in the context of the broader nuclear sector.

Interestingly, at least in some areas, there seems to be a correlation between strong center-periphery ties and greater nuclear safety. Similarly, in those regions where there have been the fewest problems in the nuclear sector, there is often less autonomy in decisionmaking power at the regional level and more central influence. That is, where Minatom has been stronger and where laboratories have maintained their relative independence from the local economy—due to a successful financing arrangement with the center—regional authorities have tended to stay out of their management decisions and facilities have, in general, been better able to protect their technology and materials. However, a dilemma of this trend is that such facilities have tended to engage in less conversion, less openness, arbitrary policymaking, and more reliance on old-style financing, all of which bode poorly for chances of downsizing and safely managing the nuclear complex. Thus, in these respects, the gradual spread of some degree of nuclear regionalism may be a necessary precondition for reform in the nuclear sector as well as the development of new local industries in their place. But democratic institutions need to become more developed and the media more independent before these conditions are likely to have much of a positive impact in the areas of nuclear safety, conversion, and downsizing.

In terms of positive manifestations expected in the environmental sector, there is little evidence to date of increased local control leading to heightened environmental safety. Instead, impoverished regions have tended to be even less concerned with nuclear safety than central authorities, particularly when nuclear power required for the local energy grid is involved. Only in a few cases—where central authorities have attempted to impose nuclear waste dumps on local regions—have local officials and the population engaged in serious resistance based on environmental grounds and improved transparency in Russia's nuclear decisionmaking.

Perhaps the greatest impediment to long-term autonomy of the regions in the nuclear sector is simply the lack of trained personnel at the local level. Technical questions requiring specialized analysis are difficult to resolve in the context of uninformed local authorities. At present, the weakness of the non-governmental sector in general and the relative absence of active and informed NGOs in particular make nuclear decisionmaking at the local level at best problematic.

Similarly the need for national controls over exports of nuclear material make central decisionmaking desirable in many instances, given the incentives within the local context of many facilities to export nuclear assets without regard to nonproliferation controls. Thus, a culture of nonproliferation needs to filter down to the local level (or be established through local educational programs) before responsible decisionmaking can occur.

In sum, the results of this research effort show that greater understanding of specific regional conditions is critical to the maintenance of nuclear safety and security in Russia. The transition to a more stable and responsible decisionmaking structure throughout the nuclear complex—with some necessary redistribution of inputs to the local level—has not yet occurred. Instead, the evidence today points to an *ad hoc* devolution of power and withdrawal of resources that have created a dangerous mix at a number of facilities. Reforms under President Putin have made some improvements, although primarily in the area of law enforcement rather than in overall structural reforms. Putin's decision to rein in local governors has been only partially effective and has come without adequate attention to improving national oversight and the local resource base. Future efforts will be needed to ensure that regional economic improvements move forward so that a nuclear safety culture can be established and physical protection of facilities can be improved and ensured. Better-informed local officials and more active local NGOs will be necessary to make this happen. In this regard, the role of the federal government in making nuclear regionalism a positive rather than a negative phenomenon will be important, as will the role of foreign assistance providers in helping to inform local regions about nonproliferation and nuclear safety requirements that must come hand-in-hand with responsible nuclear custodianship. Moscow-based and international NGOs can play an important supporting role in these efforts.

This process will take time, but it will be promoted by the influence of Internet communications and the increasingly rapid flow of information within Russia, as well as between Western countries and Russia's regions. There is a real opportunity to promote change within the Russian nuclear complex in the context of G-8's "10 plus 10 over 10" initiative. More attention by outside organizations to the regional dimension of the nuclear problem in Russia should accompany this effort, serving to enhance local capacities and knowledge. With such incentives, increased local democracy and autonomy in the nuclear sector could go hand-in-hand with an increasing capability of regional authorities to manage their regions' post-Soviet nuclear inheritances in a responsible manner.

PART I
FEDERAL NUCLEAR AGENCIES AND THE REGIONS

Chapter 2

Center-Periphery Relations and Russia's Nuclear Infrastructure

Dmitry Evstafiev and Vladimir A. Orlov

Throughout the 1990s, ties between the Russian government and the constituent parts of the federation (*sub'yekti federatsii*) were extremely fluid. Relations evolved from the Kremlin's early readiness to concede regional political offices "as much sovereignty as they could swallow," to the codification of the Federation Treaty and *ad hoc* negotiation of asymmetrical agreements for the division of authority between the center and federal constituencies. The latter phase was dramatically reversed at the onset of Vladimir Putin's presidency in 2000. Since then, center-periphery relations have been defined by federal efforts to rein in centrifugal trends and improve federal governance across the country. This cycle of centralization-decentralization-re-centralization has unfolded along both formal legal and informal political dimensions, and as a result, regionalism in Russia has turned on issues of political control and policymaking autonomy, as opposed to outright separatism.[1]

The early weakening of Russia's federal authority and ensuing contest for policymaking autonomy both directly and indirectly affected governance of the nuclear sector. On the one hand, it is clear that even during the period of extensive decentralization, ownership of the nuclear weapons infrastructure remained a "sacred cow" and beyond the scope of regional demands. Accordingly, nuclear separatism has remained a moot issue in both legal and practical terms. However, issues related to control over the commercial nuclear complex have not been insulated from legal and political tussles between federal and regional authorities. "Control" in this chapter refers not to direct operational control over nuclear facilities, but instead to power or influence over financial flows, contracts, and other policies affecting nuclear facilities. In a break with Soviet practice, formal

---

[1] With the exception of Chechnya, during the course of this contradiction-filled period, none of Russia's federation constituencies exhibited any real intention to secede. While several constituencies made declarations of their intent to do so, in fact, they had "economic" sovereignty in mind—in other words, establishing the most favorable regime for a particular region. During the period in which Yeltsin was weakened, such blackmail had a chance of being successful. The pressure of regional elites on the federal center caused concern that the federation would undergo "semi-decay."

jurisdiction over the nuclear complex has been divided into concentric circles of authority among federal, regional, and local offices. The different levels of administration wield both independent and overlapping mandates to supervise the nuclear complex. The resulting structure is a confusing hybrid of formal hierarchical legacies and new, segmented authority among the constituent members of the Russian Federation. These blurred lines of formal authority over different aspects of the nuclear infrastructure have presented opportunities for regional leaders to exercise *de facto* policymaking autonomy from Moscow.

This chapter provides an overview of the formal and informal jurisdiction over Russia's nuclear complex. It explores the political and economic relationship between center and region in the nuclear sphere, as opposed to regulatory oversight and relations between regional administrations and specific facilities that are reviewed elsewhere in this volume. The first section reviews the legal division of responsibility, summarizing the *matryoshka* structure of governance that has set the parameters for nuclear regionalism in contemporary Russia. The second section describes the informal political character of center-periphery bargaining on nuclear issues that has emerged amid the uncertainty in the formal legal structure. The third part examines the respective behavioral motives that animate "nuclear regionalism" in Russia's *matryoshka* federal structure. The conclusion analyzes the prospects for cooperation among central and regional offices in the nuclear sphere.

The Legal Context

The formal impetus for "nuclear regionalism" initially came from the "Agreement on Military Issues between the Government of the Russian Federation and the Government of the Republic of Tatarstan" that emerged out of the highly controversial bilateral treaty signed between Moscow and Kazan in 1994.[2] The agreement embodied conflicting tendencies that obfuscated legal control over the nuclear assets located on the territory of the republic. On the one hand, it formally acknowledged federal jurisdiction over military issues. On the other hand, Article 3 stipulated that Tatarstan "shall not be used for testing weapons of mass destruction, including nuclear weapons." Similarly, Article 5 specified that transportation of nuclear weapons across the territory of the republic "shall be implemented only with the expressed notification and approval of the Government of the Republic of Tatarstan." As part of its sovereign jurisdiction, Kazan also laid claim to *de facto* control over the nuclear weapons and commercial infrastructure located on its territory. Specific nuclear assets were governed by the principle of "dual

---

[2] Agreement between the Government of the Russian Federation and the Government of the Republic of Tatarstan on Military Issues, March 5, 1994, <http://www.tatar.ru/append50.html>. See also Treaty on Delineating Subjects of Authority and Mutual Delegation of Authority between Organs of State Power of the Russian Federation and Organs of State Power of the Republic of Tatarstan, February 15, 1994, <http://www.kcn.ru/tat_ru/politics/pan_for/wbi_28.htm>.

subordination" to republican and federal authority. Although the Tatar government was careful not to exploit this legal confusion and officially acknowledged Moscow's exclusive jurisdiction over nuclear-related issues, the agreement nonetheless set a formal precedent. Civic movements and interest groups in the republic, for example, seized upon the legal mandate to lobby for federal restrictions on the nuclear-related activity of government agencies within the republic. In particular, the Anti-Nuclear Society of Tatarstan held a conference, "On the Paths to Spiritual-Ecological Civilization, Problems of the Anti-Nuclear Movement, Nuclear Power, and Industry," in June 2000. According to program documents, the conference endorsed a petition for laws that "prohibit the transportation, importation, and disposal of radioactive waste and material within the territory of Tatarstan."[3]

Notwithstanding the legal ambiguity, an "Agreement on Military Issues" between Russia and Tatarstan was signed by the Kremlin to avoid serious, violent confrontation with the republic. No other treaty between Moscow and a constituent republic included an additional protocol on military issues (although nearly all regulate issues related to enterprises of the military-industrial complex). After this episode, the Kremlin made a conscious effort to retain federal ownership of nuclear materials, and to minimize regional intrusion into the formulation of nuclear policy. This commitment was reflected in the 1994 presidential edict "On Initial Measures to Modernize the Nuclear Materials Control and Accounting System," that explicitly vested authority in the federal government to oversee reform of the nuclear sphere. The edict omitted reference to "constituencies of the federation" or their legal and political equivalents, as well as implied the complete and unconditional subordination of nuclear issues (including financial matters) to federal agencies.[4]

It was only a year later, in November 1995, that President Yeltsin reversed course and signed the federal law "On the Use of Atomic Energy" that reintroduced legal ambiguity.[5] This law specified a category of "shared power" between the federal center and constituent parts in the nuclear sphere. Article 5 designated as federal property "all nuclear materials, radioactive waste containing nuclear materials, military-related nuclear installations, radiation sources, and storage facilities," while simultaneously noting that "non-military nuclear installations and storage facilities are federal property, unless otherwise stipulated by law."[6] In addition, Article 10 stipulated that federal and regional authorities "share power" to:

---

[3] "Nuzhen goskontrol nad syrevymi resursami," *Tatarskiye kraya*, No. 30, July 24, 2000.

[4] Presidential Edict No. 1923, September 15, 1994, "On Initial Measures to Modernize the Nuclear Materials Control and Accounting System," *Rossiyskaya gazeta*, September 21, 1994.

[5] Russian federal law "On the Use of Atomic Energy," *Sobraniye zakonodatelstva RF*, November 27, 1995, No. 48, article 4552.

[6] The law does acknowledge that "non-military radiation sources, radioactive substances, and radioactive wastes that do not contain nuclear materials may be federal property,

- determine the location of defense-related, federally owned, or federally or inter-regionally significant nuclear installations, radiation sources, and storage facilities;
- conduct state environmental impact studies of designs and other documents in the field of nuclear power use;
- provide for the protection of citizens during the use of nuclear power;
- provide for the safety and preservation of the environment during the use of nuclear power;
- take measures to eliminate the consequences of accidents that occur during the use of nuclear power;
- provide for training of specialists in the area of nuclear energy use, including the training of specialists in the use of nuclear installations, radiation sources, nuclear materials, and radioactive substances; and
- formulate and implement complex socio-economic development programs and environmental safety programs for the territory in which the nuclear power-related facility is located.

The delineation of "shared power" marked an unprecedented attempt by the federal center to transfer to regional administrations part of the financial burden of supporting economic and social reforms in the nuclear sector. This paved the way for regional constituencies to participate directly in final deliberations on the location of nuclear facilities across Russia. While this concession diminished the center's exclusive authority in the nuclear sphere, it was extended as part of a strategy to co-opt regional responsibility for redressing social and environmental concerns linked to reform of the nuclear sector. It was important to codify this in federal legislation, since several regions contaminated by radioactivity (including Sverdlovsk Oblast) had already adopted local laws guaranteeing the safety of the population. The prospects for extending this legislation also were discussed by administrations in the Leningrad Oblast and the city of St. Petersburg.[7]

In addition, the law "On the Use of Atomic Energy" delineated areas of independent jurisdiction for regional agencies. In particular, regional administrations were authorized to:

- supervise radiation sources, storage facilities, and radioactive substances considered the property of Russian Federation constituencies;

---

federation constituency property, and municipal property, as provided by law." Moreover, "property rights to these objects are formulated on the basis of evidence provided by the Russian Federation Government in the procedure established by it." Thus, a distinction was made between nuclear materials and other radioactive materials, which is fair.

[7] This law was adopted by the State Duma on February 21, 1996, approved by the Federation Council on March 20, 1996, and signed by President Yeltsin on April 3, 1996. Valentina Romanova, "Atomnaya energiya: kto otvetit, yesli chto?" *Chas Pik*, No. 42, October 25, 2000.

- implement within their territorial jurisdiction measures to ensure the safety of nuclear installations, radiation sources, and storage facilities within the limits prescribed by Russian federal law;
- develop, in coordination with federal programs concerning the use of nuclear energy, appropriate republican and regional (territorial) programs;
- establish procedures, in conjunction with public associations and citizens groups, for discussing issues related to the use of nuclear energy on their territory;
- establish procedures for determining the location and construction within their territorial jurisdictions of nuclear installations, radiation sources, and storage facilities considered the property of Russian Federation constituencies; removing such sites from operation; and storing radioactive wastes;
- resolve issues related to the protection of citizens and the environment from radioactivity that exceed the limits established by norms and regulations of the nuclear energy sphere;
- ensure the radiation safety of the populace and protection of the environment within their territorial jurisdictions, as well as the preparations of organizations and citizens to act in case of an accident at a nuclear power facility;
- implement procedures for controlling and accounting for radioactive materials within their territorial jurisdictions, under the state system of radioactive materials control and accounting;
- implement physical protection of nuclear materials and of nuclear installations, radiation sources, storage facilities, and radioactive substances considered the property of Russian Federation constituencies;
- resolve other problems in the area of nuclear energy use within the limits set by Russian federal legislation.

In contrast to areas of "joint jurisdiction," independent regional responsibilities were defined broadly, permitting—given sufficient political will—regional administrations to weigh in on practically any issue concerning the nuclear activities conducted on their territories. The designation of nuclear facilities with "regional importance" ceded federal subjects authority to claim partial ownership of the nuclear infrastructure that subsequently became a subject of intense controversy between federal and regional offices. At the same time, the closed cities, formally known as closed administrative-territorial formations (ZATOs), were excluded from regional jurisdiction.

The law also stipulated local jurisdiction in the nuclear sphere. Article 12, for example, empowered municipal authorities to:

- deliberate and resolve questions related to the location of nuclear installations, radiation sources, and storage facilities within their territorial jurisdiction;
- decide on the location and construction of radiation sources of local importance within their territorial jurisdiction;

- participate in the generation of environmental impact studies of the plans for sites that use nuclear energy and of nuclear installations, radiation sources, and storage facilities slated for construction within their territorial jurisdiction;
- propose land within their territorial jurisdiction for placement of nuclear installations, radiation sources, and storage facilities in accordance with Russian federal legislation;
- inform local constituencies through mass media of the radiation situation within their territorial jurisdiction;
- develop and adopt measures to protect citizens and their personal property, to minimize damages, to restore normal activities of organizations in case of an accident at a nuclear facility, to inform the public in a timely manner of the threat of radiation exposure and levels of radioactive contamination of the environment and agricultural production.

The law followed a general political course of placing regional governments in opposition to municipal administrations that was consistent with Moscow's efforts to counter trends towards regionalism from 1997 to 1999. During this period, however, there were no overt nuclear controversies among constituent administrations, as nuclear issues typically did not offer leverage for consolidating local power in opposition to regional "barons." Nonetheless, the legal provisions enabled municipal leaders to foist themselves into debates over reform of the nuclear power sector and military activities, as well as to champion related political issues in efforts to garner local public support.

The result was the creation of a *matroyshka* federal structure for supervising the nuclear sector, as depicted in Figure 2.1.

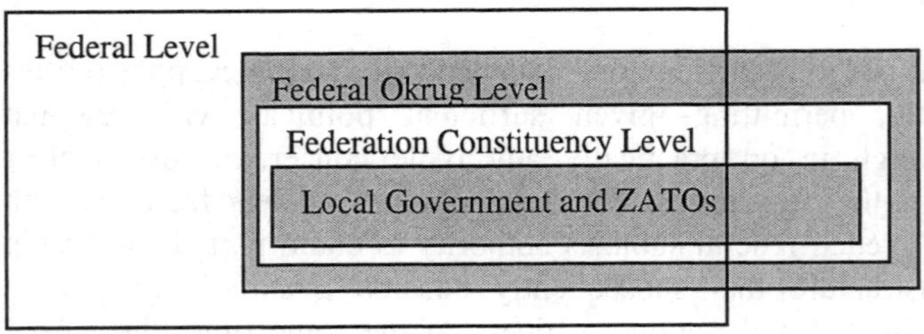

Figure 2.1 Russia's *Matryoshka* Federalism

This system was a by-product of the efforts to both strengthen the regions relative to federal power, and strengthen the government relative to the power of the president. Both trends reached a climax with President Yeltsin's efforts to co-opt support for his re-election in 1996. It was the very concessions made by the center to the federal constituencies in 1995, and not the populist sparks of the early 1990s, that created the conditions conducive for the nuclear regionalism that emerged during the late Yeltsin era.

The primary regional objective, however, was to gain access to federal financing to reform the Russian nuclear sphere. In this sense, the federal center took steps to meet the regions part-way. According to Article 3 of the 1996 federal law, "On financing especially radiation-hazardous and nuclear-hazardous production and facilities":

> Organizations and military units included in the list of especially dangerous facilities may make deductions/allocations for the creation of centralized funds included in the production costs (labor/service costs), regardless of source of payment, at the following rates: ...one percent of the production costs of goods sold, in order to finance measures for social security for the population residing in territories adjacent to especially hazardous facilities, which are determined according to the procedure established by the Russian government, and in order to finance the development of social infrastructure in these territories.[8]

In essence, the law strengthened the legal basis for regional access to federal funding in the nuclear sphere. At the same time, Moscow's position on other aspects of reform of the nuclear complex remained very strict. For example, the 1996 government decree "On approving 'The statute on the procedure for exporting and importing nuclear materials, equipment, special non-nuclear materials, and related technology'," precluded federal subjects from participating in the nuclear decision-making process.[9] Ironically, this statute was approved at a time when President Yeltsin was in great need of electoral support from regional leaders. Regional authorities traditionally demonstrated commercial interest on issues related to high technology exports and attempted to mark their presence in this sphere.

In sum, the legal basis for interaction in the nuclear sphere between Moscow and the regions significantly evolved following the Soviet collapse. The federal center initially succeeded at retaining exclusive authority over the entire nuclear military and civilian infrastructure. This position eroded somewhat due to the perceived weakness of the federal center. Yet, the long-term initiative remained firmly in the grasp of the federal center, as the 1993 Russian constitution empowered the government to grant tactical concessions to contain regionalism in the nuclear sphere. It also created the institutional context for leveling potentially dangerous aspects of nuclear decentralization after President Putin spearheaded his federal reforms.

---

[8] Russian federal law "On financing especially radiation-hazardous and nuclear-hazardous production and facilities," *Sobraniye zakonodatelstva RF*, April 8, 1996, No. 15, article 1552.

[9] Government Decree No. 574 "On approving 'The statute on the procedure for exporting and importing nuclear materials, equipment, special non-nuclear materials and related technology'," May 8, 1996.

The Phases of Russian Nuclear Regionalism

Although the legal framework set certain limits on regional control over Russia's nuclear infrastructure, deep political disagreements accompanied the interpretation of laws, edicts, and norms promulgated at the federal center. This situation was partly the result of the *de facto* division of authority among agencies within the Russian government. The "regionals" (i.e., the informal coalition of elected regional leaders) emerged from within this system between 1996 and 1998 as the system acquired its basic contours. The authority of the "regionals" in the nuclear sector was primarily informal, despite the legal parameters briefly reviewed above, and cut across political, economic, fiscal, environmental, social, and nuclear safety issues. In this regard, "nuclear regionalism" pertained to the roles and places of the federal center and federal constituencies at different stages in the process of reforming the Russian nuclear complex. At its core, this concerned questions about how the changes in the alignment of forces between the participants in the process took shape. In this sense, the decentralization of nuclear authority unfolded as a limited partnership between the federal center and regional administrations to cope with the country's socio-economic problems. Accordingly, this issue was inseparable from the general state of political and economic affairs in the country.

Stage 1: 1991–93

The first stage in the creation of the new Russian state was marked by the emergence of a political diarchy. At the sub-federal level, elites managed to accelerate their consolidation of power in the republics, while in the krays and oblasts their power bases remained amorphous. On nuclear issues, the primary concern was related to nuclear safety and transparency of scientific and industrial activity occurring at nuclear complex enterprises. There was a dearth of full-scale legal mechanisms to regulate interaction between the center and regions on nuclear reform. Consequently, the period was marked by intense public and elite populism on issues pertaining to the military and commercial sectors of the nuclear industry.

Stage 2: 1993–95

The second stage was marked by the consolidation of presidential power and liquidation of political divisions at the federal level, as well as the expansion of the economic authority of regional elites. Regional leaders began to consolidate power in the krays, oblasts, and national republics. There was a gradual weakening of political regionalism in favor of economic regionalism, and the formation of inter-regional economic cooperation associations. The first legal conflicts concerning nuclear infrastructure began to take place, as the fully fledged legal basis for relations was only in its nascent stages. This also introduced center-periphery conflicts over control of property in the potentially profitable military-industrial complex and nuclear infrastructure. The populist motives that animated regional activism in the nuclear complex quickly subsided, as regional elites began to view the local nuclear infrastructure as an economic asset. Accordingly, regional leaders gradually became allies of the federal elites in countering local anti-nuclear and environmental movements.

Stage 3: 1995–99

The third stage was defined by a relatively stable balance of power between the federal center and regional elites. There was widespread appreciation for the mutual benefits that could be potentially derived from reforming the local nuclear infrastructure. On the federal level, legal norms establishing control were codified that virtually precluded regional participation in nuclear-related decisionmaking. Regional leaders, however, continued to exploit issues related to nuclear reform as part of a propaganda campaign aimed at pressuring the federal government for greater autonomy. It was during this period that international (primarily U.S.) financial assistance for reducing the risk of nuclear proliferation became a significant factor in Russian nuclear politics, spawning renewed competition between the center and regions for control over the industry. This factor underlined the financial attractiveness of the nuclear complex. "Slicing the pie" of foreign aid provoked new contradictions between the center and regions.

Stage 4: 1999–present

Putin's first term was characterized by the strengthening of the center's administrative and political control over regional elites. There was a gradual reduction of political regionalism and confirmation of the federal center's leading role in continuing Russian reforms. This stage was marked by the reduced salience of nuclear issues in the center-periphery political agenda but increased importance of these issues in terms of shaping federal economic relations. While regional elites acknowledged the authority of the federal center to determine the substance and methodology for managing nuclear reforms, they continued to clutch to their claims to overseeing local commercial activities in the nuclear sector.

Regional Elites and Problems within the Nuclear Complex

There was a paradox concerning the status, fate, and economic use of the components of the Russian nuclear complex. On one hand, the legal status of the Russian nuclear complex within the framework of center-regional relations was determined in the mid-1990s. It was never explicitly recognized as a free-standing issue in the subsequent debates between the center and regions. On the other hand, issues related to control over the nuclear infrastructure featured prominently in political battles over the redistribution of power between the center and those regions with nuclear facilities on their territories. At times, regional actors attempted to insert themselves into the debates over reform of the nuclear complex. These efforts were motivated by a variety of reasons.

One motive for assertiveness on nuclear issues was linked to the popularity of specific regional leaders. The quintessential example was Krasnoyarsk Kray Governor Aleksandr Lebed's announcement concerning the possibility of subordinating the Strategic Rocket Forces divisions based in the kray to regional authorities. His rationale for issuing the threat was that the federal government failed to meet its fiscal obligations to these divisions. This proposal overtly contradicted federal law. Yet, Lebed did not undertake the initiative to increase his influence in Moscow; rather, he did so to provoke the Kremlin and to paint himself as the "defender" of the region's interests. On the regional level, where the insufficient financing of military units (and not just those related to the SRF) aggravated local social problems, this announcement had the immediate effect of bolstering his local popular standing.

A second motive for regional activism was associated with efforts by the governors to bolster their prestige as federal-level politicians. Regional leaders dearly valued ties—even if only from the vantage point of propaganda—to nuclear issues. They believed that by addressing such issues they would be taken seriously as politicians on the national stage—especially by their regional constituents. This was especially characteristic of politicians who viewed their gubernatorial posts as springboards for national office. Besides Aleksandr Lebed, other governors that adopted this posture included Dmitry Ayatskov (Saratov Oblast), who was the most active of all regional leaders in addressing nuclear issues. Similarly, Sverdlovsk Oblast Governor Eduard Rossel actively interacted with the Russian Nuclear Society and even facilitated the convening in September 1997 of its annual conference in Yekaterinburg. His opening address at the forum was not strictly a propaganda move to bolster his image. Rather, during the speech he advocated installing a fourth unit at the Beloyarsk NPP and proposed creating a public corporation to handle extra-budgetary financing of the construction. At Rossel's initiative, a symposium entitled "Atomic Urals" was convened that reviewed issues related to the clean-up of nuclear enterprises in the Urals and the results of a government program to deal with health and environmental problems attendant to activities at the Mayak Production Association. As a result, Rossel foisted himself into the national debate over managing the nuclear legacy of the Soviet Union and sought to exert control over the activities of federal structures.[10]

Similar goals were pursued by Primorskiy Kray Governor Yevgeniy Nazdratenko, who otherwise did not have an affinity for playing the nuclear card in his dealings with Moscow. In particular, the governor issued a decree on December 7, 2000 creating an inter-agency commission to conduct a thorough study of facilities located within the kray that possessed nuclear materials, radioactive substances, explosive substances, toxic substances, and pathogens. Sensing that relations with the Kremlin were deteriorating, Nazdratenko decided to flaunt his credentials as a national politician to the local population and military. This decree stipulated that inspections take place at facilities assigned to the Pacific Fleet and Far East Military Okrug in order to prevent possible environmental disasters

[10] Sverdlovsk Oblast Governor's Press Service, Press release, September 15–17, 1997.

related to accidental releases of rocket fuel components and radioactive waste.[11] In the process, Nazdratenko legally sanctioned indirect regional civilian control over military activity, and joined efforts to solve state tasks in the area of environmental security.

As noted earlier, the maneuvering by the center and regions in the nuclear context generally boiled down to the following: the center was not interested in integrating regional elites into the process of restructuring the Russian nuclear complex, but regional elites found ways to become involved. Even as influential a political insider as Moscow Mayor Yuriy Luzhkov tried to intimate himself into the system of relations involving the nuclear sphere, introducing methods for exchanging debts of the All-Russian Scientific Research Institute for Experimental Physics (VNIIEF) in Sarov for Moscow apartments. This scheme had no commercial component for the Moscow government or for Minatom officials, but demonstrated Luzhkov's role in resolving issues in the national security sphere.[12]

In addition to issues of prestige, Russian regional leaders were driven by practical considerations to participate actively in deliberations pertaining to nuclear issues. This included ensuring the completion of construction of nuclear power facilities, integrating them into respective regional economic systems, and regulating social tensions that may arise at local nuclear facilities. Also of great importance are the environmental consequences of activity in the nuclear sphere. Murmansk Governor Yuriy Yevdokimov, for example, addressed politically sensitive issues related to the financing of nuclear submarine dismantlement within his oblast, and the state of nuclear safety primarily out of necessity to retain his post. To demonstrate his concern for issues important to the populace, and to prevent the occurrence of real environmental catastrophes within the oblast, the governor was obliged to make a point of discussing nuclear problems and offering substantive recommendations in the national decisionmaking process.

A fourth impetus for regional assertiveness on nuclear issues has been related to the prospects for gaining access to additional economic resources. This motive received the least amount of attention by the Russian public. However, it always lurked in the "shadow" of the dialogue between the center and regions. This economic dimension was most conspicuous in the campaigns by Governors Ayatskov and Lebed, who lobbied Minatom directly for commercial programs. The same concerns held for local projects as well. In particular, in April 2000, during the course of discussion of the primary direction of reforming RAO United Energy System, Saratov Oblast Governor Ayatskov announced his preparedness to facilitate the reclassification of the Balakovo NPP to a heating plant (according to the way the energy is obtained) and facilitate its transfer to a property of the region.

---

[11] "V Primorye sozdana kommissiya po obsledovaniyu radioaktivnykh obyektov," *RosBiznesKonsaltiing*, December 7, 2000.

[12] Olga Zoguskina, "Dolgi Sarovu—kvartirami v Moskve," December 10, 1997, <http://www.portal.sarov.ru/curier>.

Regional leaders also systematically acted as lobbyists for the interests of nuclear fuel cycle facilities within their territories out of commercial interest. For example, Leningrad Oblast Governor Vladimir Yakovlev lobbied for a project to create a facility for long-term handling of nuclear waste from aging Northern Fleet nuclear submarines that was proposed by British Nuclear Fuels, Ltd. In this instance, Yakovlev acted not in his capacity as governor, but as president of the Northwest Regional Association, thus raising his political stature. In large part due to the initiative of St. Petersburg officials, new equipment for ensuring the safety of the nuclear reactor of the transport ship *Sevmorput* was made at city factories, within the framework of a Russo-Swedish-British project.[13] In general, the oblast leadership was actively engaged on issues related to strengthening nuclear safety at Leningrad NPP. In particular, the regional administration supported various public measures related to improving nuclear safety, including the convening of an international conference in June 2001 on a "Thorough Assessment of Safety of the 2nd Reactor Unit at Leningrad NPP."[14]

However, not all regional leaders regarded the nuclear complex and its elements as intrinsically profitable to their territories. The situation varied both across regions and facilities. For example, officials in Nizhniy Novgorod and Tomsk Oblasts tended to associate the presence of closed cities (Sarov, Seversk) with an expected infusion of economic wealth into the region. In contrast, officials in Chelyabinsk Oblast regarded the closed city of Snezhinsk as an administrative headache and a liability to the regional budget. In a number of cases, regional leaders viewed local nuclear facilities as a net burden that should be excised from the regional budget altogether. Given the *carte blanche* extended to other sectors of the regional economy—such as the local electrical power grid, the petroleum industry, and import-export operations—there was little commercial value to supporting local assets associated with the nuclear fuel cycle.

Finally, regional leaders are motivated to address nuclear issues as a part of political gambits aimed at preempting or co-opting local non-governmental environmental movements. As a rule, regional environmental organizations are negatively disposed toward regional elites. They periodically accuse local authorities of ignoring environmental issues in favor of commercial interests. On occasion, this required that regional leaders react with decisive and robust propaganda campaigns, as was the case for the Murmansk Oblast administration. Although it had the trappings of a typical debate on environmental issues, in fact it was about control over financial resources disbursed by the European Union's Technical Assistance to the CIS (TACIS) program (beginning in 1994) for implementing the program NUCRUS 95410.[15] By framing the issue in terms of the potential commercial profit to the region, local leaders were able to sideline the

---

[13] *Ekonomika i vremya*, No. 32, August 20, 2001.

[14] Igor Lisochkin, "Proyekt blestyashche zavershen, sotrudnichestvo prodolzhayetsya," *Sankt-Peterburgskiye vedomosti*, No. 119, June 29, 2001.

[15] The ultimate goal of the program was the project for long-term storage of radioactive waste for the northwest region of Russia.

environmentalists and manipulate public support on the issue.[16] The regional leader that came under the most pressure from environmentalists was Sverdlovsk Governor Eduard Rossel. This was due, in part, to the co-optation of environmental concerns raised by the construction of long-term radioactive waste storage facilities in the oblast by regional political organizations opposed to Rossel. The so-called "May Movement" (a left-leaning parliamentary and public lobby), in particular, was very influential in the Urals region.[17] The criticism of Rossel was grounded in the fact that he worked closely with Minatom officials on a wide range of issues. He was accused in the regional press of placing commercial and national political ambitions ahead of regional, social, and environmental concerns.[18] These accusations subsequently formed the crux of a meeting between May Movement representatives in the Yekaterinburg Legislative Assembly and Petr Latyshev, the presidential envoy to the Urals Federal Okrug.[19]

The conflict between the May Movement and Rossel illuminated the extent to which nuclear debates animated regional political battles. The combination of ignorance on the part of the average person and the venom espoused by environmentalists in local media, as well as intense concern over nuclear safety (a "Chernobyl syndrome" magnified by the fear of terrorist threats), could have explosive results. In addition to more reasonable accusations, the May Movement charged Rossel with "allowing underground peaceful nuclear explosions to be conducted within the oblast in collusion with Minatom leadership and without informing the public." This accusation, which experts knew to be false, preempted constructive dialogue on the merits of nuclear versus alternative energy sources for the region.

A similar conflict arose in Chelyabinsk Oblast. There, local officials endorsed construction of the South Urals NPP, despite the publicly sensitive problem of lingering territorial contamination. This elicited a harsh reaction from the local environmental movement, which tried to organize legal prosecution of proponents of continued construction, which included Mayak and the oblast administration.[20] The response of the local administration, which viewed the South Urals NPP project as strategically important for regional development, precipitated a propaganda battle in the region. Both sides used environmental arguments to influence the public and hardly touched on the economic merits of the program.[21] The Chelyabinsk Oblast administration also sought to overcome the anti-nuclear syndrome. In particular, South Urals State University regularly held "Kurchatov Readings," which were oriented towards educating students and the local

---

16 See <http://www.murman.ru>.

17 See "Thursday, March 9, 2000," *Guest Book of Socio-Ecological Union of Russia*.

18 See, for example, "Rossiyskiy atom gotov stat mirnym," *Kommersant-daily*, May 25, 1999.

19 Urals Information Bureau, March 5, 2001.

20 Karina Panova, "Yesli prorvet plotinu," *Sinegorye*, No. 35, August 29, 2001.

21 Yelena Vasilyeva, "Atomnaya stantsiya nas spaset," *Vecherniy Chelyabinsk*, No. 166, September 7, 2001.

intelligentsia on nuclear-related issues.[22] In other oblasts, such as Tomsk, regional authorities, in close collaboration with Minatom, organized educational activities aimed at reducing the popularity of anti-nuclear slogans that otherwise stymie the implementation of potentially economically profitable projects.[23]

Minatom traditionally served as an active ally of regional authorities in countering regional environmental movements. In a 1993 referendum, residents of Balakovo, Saratov Oblast, voted to ban construction of units 5 and 6 of the Balakovo NPP. In early 2001, Minatom officials announced the allocation of 245 million rubles for construction, which revealed that a basic decision had been made to begin construction in opposition to efforts at holding a local referendum. In this controversy, the oblast leadership completely supported Minatom, explaining that the situation was economically expedient.

The Union of Nuclear Energy Territories and Enterprises had a significant effect on the decision by Rostov Oblast authorities concerning construction of an NPP there. The oblast leadership eventually sided completely with Minatom on the construction issue, considering that it would be economically beneficial and would increase the political status of the oblast and its leaders. As declared in a speech by Rostov Oblast Government Chairman, I. A. Stanislavov, at a session of the Board of the Union of Nuclear Energy Territories and Enterprises, "There is no union of territories in which coal or gas power plants are located. There is only the union of territories in which nuclear power plants are located. And we must consider this in our actions." Consequently, the oblast leadership reserved the right to control the course of construction and operations at the NPP.[24]

Paradoxically, the most extensive local intrusion into decisionmaking on nuclear issues was evidenced by regional politicians who, in one way or another, cast themselves as strict federalists (such as Ayatskov of Saratov Oblast, Konstantin Titov of Samara Oblast, and Ivan Sklyarov of Nizhniy Novgorod Oblast). For example, while discussing the issue of importing spent nuclear fuel into Russia, Irkutsk Oblast Governor Boris Govorin underlined his stand against importing spent fuel into, or transporting it across, his oblast. This was done to consolidate support, in the face of upcoming gubernatorial elections, from the maximum number of public organizations and associations in the oblast.[25] Also pitted against the Kremlin in the debate was Novgorod Oblast Governor Mikhail Prussak, considered an ardent supporter of strengthening central government and liquidating the vestiges of confederalism in Russian state power.

At the same time, regional leaders traditionally associated with broader interpretations of the scope of political regionalism (such as Tatar President Mentimer Shaymiyev, Bashkortostan President Murtaza Rakhimov, and Chuvash

---

[22] Mikhail Fonotov, "Ot radiofobii spacayut tolko znaniya," *Chelyabinskiy rabochiy*, No. 63, April 5, 2001.

[23] N. Chernyshova, "SKhK prinimayet mery protivodeystviya teraktam," *Vse dlya vas*, No. 252, April 12, 2001.

[24] "Tri problemy Rostovskogo regiona," *Priazovskiy kray*, No. 27, July 5, 2001.

[25] *Argumenty i fakty*, June 21, 2001.

President Nikolay Federov) have not participated, or have participated minimally, in discussions of nuclear infrastructure reforms. Both Shaymiyev and Rakhimov were conspicuously silent during the legislative debate over the spent nuclear fuel amendments. And yet politicians of their kind did not decline to engage in direct debate with the center on other less significant issues. Under Putin, the more active "regionals" faced the greatest difficulties, compelled to endure persistent multi-level "haggling" with the Kremlin that came at great cost to their political stature. In this context, it became politically risky for them to address sensitive topics, such as those pertaining to the nuclear sphere.

While independent-minded regional executives, such as Shaymiyev, demonstrated restraint on nuclear issues, local parliamentarians persisted with direct challenges to federal authority. Places have been taken by regional elites who speak through local legislatures. For example, on March 28, 2001, the deputies of the State Council of Tatarstan adopted an open letter to the president, cabinet, and Federal Assembly of Russia demanding that the ban on nuclear waste imports into and temporary storage within Russia be upheld. The letter also noted that the role of federal constituencies should not be underestimated: "The State Council of the Republic of Tatarstan is firmly convinced that the resolution of acute environmental problems, including provisions for nuclear and radiation safety, directly affects the interests and fate of the people and territories, is possible only given a comprehensive approach, based on consideration of the opinions of all constituencies of the Russian Federation."[26] Tatarstan State Council Speaker Farid Mukhametshin explained that "(negative) consequences of importing and storing nuclear wastes in Russia could be much greater than the benefits the country may receive in the first stage," and that Russia must not "place economic benefit before the environmental safety...of the population."

Conclusion: Prospects for Center-Periphery Cooperation on Nuclear Issues

It is common to assume that under President Putin's policy of recentralization, nuclear regionalism became increasingly a moot issue. This, however, has not been the case. This is due in part to the two faces of decentralization. On the one hand, nuclear regionalism has had a negative impact, raising the specter of a general breakdown in control over the Russian nuclear complex and the threat of a "tug of war" between the regions themselves over control. This threat is less clearly defined today than it was in earlier stages of the evolution of the Russian state. On

---

[26] "Open Letter of the State Council of the Republic of Tatarstan to President of the Russian Federation V. V. Putin, Chair of the Federation Council of the Russian Federal Assembly Ye. S. Stroyev, [and] Chair of the State Duma of the Russian Federal Assembly G. N. Seleznev" on the issue of the unacceptability of adopting laws permitting the import into Russian Federation territory of radioactive materials of foreign states, including spent nuclear fuel (irradiated fuel assemblies of nuclear reactors), adopted by RT State Council March 28, 2001, <http://www.rt-online.ru>.

the other hand, there has been a more constructive dimension to the devolution of authority with respect to control over the nuclear complex. In particular, the center and regions have experienced different stages in their relationship concerning the nuclear complex over the last decade, and have learned to be partners—in some cases sharing the burden of responsibility and sharing the economic benefits of the development of the nuclear power industry. In this respect, political decentralization may have facilitated more robust control over the complex. Yet, it would be overly hasty to conclude that the risks associated with nuclear regionalism are a thing of the past. Indeed, current circumstances in Russia demonstrate that the issue of cooperation between the center and the regions in the nuclear sector has not only not dropped off the agenda, but is one of the more timely and urgent issues.

Although the impetus for regional assertiveness has been tied to factors that are deeper and that predate Putin's efforts as recentralization, there is no reason to expect that it will manifest itself in a chaotic manner. Despite having established themselves as political actors independent of the Kremlin during the election campaign of 1996–97, and having reached the peak of their influence in winter 1998 to spring 1999, regional elites were consistently unable to formulate national political and economic policies to manipulate the weakness of the center to their advantage in the nuclear sphere. As a result, with the change of the political climate in Moscow, the political advantage of the regions was first nullified, and then a massive campaign began to remove political and economic authority from regional authorities. This was formally manifested in the creation of seven federal districts and in the appointment of the presidential envoys.

With the increasing economic attractiveness of a number of nuclear fuel cycle enterprises on the one hand, and the population's greater sensitivity since 1999 and especially after September 11, 2001, to the risk of nuclear terrorism, on the other, regional elites are paying greater attention to potential nuclear "trump cards." Reform of the nuclear industry remains unfinished, and Minatom—by and large—retains its unique role as an empire-fortress. Yet, while opportunities for regional elites to use their trump cards may have decreased, it is important to note that they have not been lost altogether.

Chapter 3

Russian Nuclear Regionalism at the Federal Center

Nikolai Sokov and Adam N. Stulberg

President Vladimir Putin's sweeping federal reforms in 2000 marked a *volte face* to the political decentralization that had gathered momentum in Russia during the troubled later years of Boris Yeltsin's administration. In reaction to the politically expedient asymmetrical deals cut with independent-minded governors and the *ad hoc* devolution of jurisdiction under President Yeltsin, the Putin team arrived in office determined to take back federal power from the regions and restore the "vertical dimension" of control.[1] The new administration confronted the challenge head on, exploiting Putin's considerable popularity and the fluidity of a protracted election season in the provinces to advance a series of policies and institutional reforms that fundamentally recaptured the initiative and shook up center-periphery relations. With new structures in place, Moscow moved decisively to improve coordination of all agencies and facilities under federal jurisdiction that were spread across Russia, including those charged with implementing the nuclear policies of the Ministry of Atomic Energy (Minatom), the Ministry of Defense, and the Russian Shipbuilding Agency (which assumed responsibility for the nuclear shipyards in 1999). By the end of the first year of Putin's presidency—even intimation of nuclear separatism, in the spirit of Governor Aleksandr Lebed's oft-cited 1998 threat to commandeer the Strategic Rocket Forces division deployed in his region—was for all practical purposes a moot issue.

Notwithstanding the early achievements of Putin's federal initiatives, key aspects of both the civilian and military dimensions of nuclear policymaking remained territorially decentralized. As noted in Chapter 1 and detailed in Part II of this volume, regional administrations continued to employ their own set of policy tools to govern local matters in ways that obstructed and, in some cases, eased implementation of federal policies. Moscow's persistent failure to deliver promised support to regional sub-units perpetuated control problems that continued to both frustrate and expedite nuclear safety and security at local facilities, as well as to challenge the credibility of Russia's international arms control commitments. But

---

[1] Throughout the chapter, "regions" refers to all units below the federal level (i.e., republics, krays, oblasts). "Regional leaders" refers to the executive heads of these units (i.e., republican presidents and governors).

this process was not uni-dimensional, with regional interventions confined exclusively to complicating the execution of federal policies. Much overlooked in the broader debate over regional influence and the devolution of political control were "how" and "to what extent"—within the predicament of "weak center-weaker regions"—Russia's provincial leaders continued to affect the formulation of policies at the national level.[2]

Indeed, Russia's regions remained important political players at the front end of national policymaking. Despite Putin's crackdown on fissiparous tendencies and restoration of executive dominance, regional leaders continued to wield procedural powers at the federal center, both directly through their representatives in the national legislature and indirectly via political parties and region-based interest groups. Having seized upon special privileges and newly revamped institutions for representing their interests, regional elites were able to mobilize select provincial concerns, lay down political markers, and forge alliances with other powerful groups at the center to constrain the general contours of federal policy.

By the end of Putin's first term in office, federal-level manifestations of Russian decentralization appeared to affect nuclear policymaking in two nuanced but distinct ways. First, newly streamlined federal institutions provided regional authorities with indirect channels to set the terms for advancing proposals for nuclear energy reform. Because federal ministries and agencies had to submit policy initiatives for review and approval to bodies where they could anticipate encountering regional dissent, they had incentives to avoid certain issues and build selective regional preferences into proposals at an early stage in the policymaking process. By preempting specific policy initiatives, otherwise preferred by Minatom or other lobbies at the federal level, provincial interests shaped the general parameters for ensuing policies. Second, new federal institutions provided venues for building alliances between organized regional interests and other highly institutionalized national interest groups, whose preferences converged on specific nuclear energy issues. These coalitions strengthened the leverage of specific regional interests in the lobbying of federal legislation and policies. Together, these new federal institutions served as fora for bolstering transparency and accountability in national policymaking. They served as mechanisms for drawing public attention to divisive policies and for mobilizing collective action at the federal and regional levels, both in support of and in opposition to future implementation of nuclear energy policies. The net effect was that these new institutions extended regional influence throughout the process of formulating federal nuclear energy policies that was more subtle and effective than otherwise revealed by formal voting and direct legislative action.

This chapter develops these themes, first by analyzing the impact of Putin's federal reforms on the representation of regional interests in national policymaking. Specific attention is devoted to moving beyond the strict vertical

---

[2] Thomas Graham, Alexander J. Motyl, and Blair Ruble, "The Challenge of Russian Reform at a Time of Uncertainty," in *Russia Initiative: Reports of the Four Task Force* (New York: Carnegie Corporation, 2001), <http://www.carnegie.org/pdf/rimaster.pdf>.

dimension to center-periphery relations by reviewing the formal institutional basis for regional input on federal policies provided by the reconstituted Federation Council and State Council. The focus then turns to analyzing three informal avenues of regional influence. Here, we examine how regional interests leveraged new authority within these federal institutions to affect two specific policies in the nuclear energy sector: 1) legislation allowing for the import of foreign-origin spent nuclear fuel; and 2) reform of the electrical power industry and national electricity monopoly, Unified Energy Systems (Rossiskoe Aktsionernoe Obshchestvo "Yedinaya Energenicheskaya Systema" or RAO EES). The chapter concludes with a discussion of the implications of the "federal face" of nuclear regionalism for the shifting politics of nuclear energy reform in Russia.

Putin's Federal Reforms

In response to the legacy of weak federal institutions and the lack of consensus on basic issues of sovereignty and jurisdiction, President Putin moved quickly upon taking office to restructure the institutions that regulated center-periphery relations in Russia. The purpose of these initial undertakings was to restore central political authority in specific spheres of federal and joint competence at the expense of regional governors. As is well known, one track was devoted to improving the effectiveness of federal mandates and harmonizing policies at the regional level.[3] The second track introduced a series of laws and decrees that were designed to alter regional influence at the federal level. The latter included the August 2000 requirement that the governors and regional parliamentary chairmen relinquish their *ex officio* membership in the Federation Council, the upper chamber of the federal assembly. Under the new law, the regions gradually became represented by two full-time delegates, one appointed by the regional executive leader and one

[3] The most widely heralded initiative in this vein was the creation in May 2000 of seven federal districts, based on existing military districts and headed by plenipotentiaries charged with overseeing the President's constitutional authority in the provinces. These envoys, five of whom were selected by Putin for their backgrounds in the military and security services, were assigned the critical tasks of strengthening vertical discipline among federal offices working in the regions and bringing "deviant" regional legislation into conformity with federal norms. This was followed by a series of legal reforms that afforded the Russian president conditional authority to remove criminally negligent governors from office and to disband regional legislatures that adopt unconstitutional laws. They also stripped the governors of powers to appoint regional police chiefs. Federal intervention at the regional level was pushed further in June 2001 when Putin established a new commission for delimiting federal and regional jurisdictions, headed by a deputy chief of the Kremlin apparatus and administered by the federal envoys. This new body was charged expressly with reviewing the bilateral treaties between the center and regions signed during the Yeltsin era, with an eye towards clarifying "who is responsible for what" and creating a single legal space across Russia. *Izvestiya*, June 27, 2001.

nominated and approved by the regional assembly.[4] In principle, this could have established a more balanced federal structure, as the Federation Council acquired a "professional" membership, leaving the governors to concentrate on regional affairs. Yet, there were certain additional features of these reforms that weakened both the upper chamber and the governors. In particular, the governors lost limited immunity from criminal prosecution and related perks that they once enjoyed as federal officials. By changing the composition of the Federation Council in this manner, Putin also effectively reduced the national stature of the governors, and restricted their direct participation in national policymaking.[5] These provisions constituted a formal constraint on the autonomy of the upper chamber, and potentially weakened legislative oversight at the federal level.

The effort to curb regional influence at the center was subsequently reinforced by formation of the State Council.[6] Established by decree to compensate the governors for the re-constitution of the Federation Council, the body was chaired by Putin and included all 89 regional executive leaders (serving on a rotating basis), as well as other federal officials and delegates appointed directly by the president. However, unlike the Federation Council, which derived standing directly from the Russian Constitution, the State Council was created strictly as a consultative organ that was intended to convene once every three months. In particular, it lacked the backing of federal law, a clear mandate, access to funding outside of the presidential apparatus, and its own staff. In addition, Putin personally retained authority to appoint members of the presidium, who were expected to rotate every six months, as well as to set the agenda for each meeting. Thus, while the State Council could offer advice on specific policy issues, the president was under no formal obligation to consider it or take action on it.

---

[4] Under this law, which came into full effect in January 2001, old members of the Federation Council are to be replaced gradually as their regional terms of office expire. New members are appointed by both regional executive and legislative leaders. The speaker of the regional assembly is responsible for nominating the legislative delegate, but nominees can be accepted from the floor of the parliament that receive at least one-third of the body's support. The delegate is selected by a secret floor vote. The executive branch delegate is appointed by the regional leader. This appointment is subject to a veto by two-thirds of the regional legislature. Both delegates are to serve for two-year terms that run concurrently with their regional executive and legislative branch appointers. In addition, regional executive and legislative bodies can recall their delegates, if they are dissatisfied with their voting behavior in the Federation Council.

[5] With the subsequent formation of the pro-Kremlin *Federatsiya* parliamentary group, comprised of two-thirds of the members of the new Federation Council and the election of Sergei Mironov (a Putin ally from St. Petersburg) as its new speaker in December 2001, the Russian president quickly parlayed these institutional revisions into a springboard for the emergence of an internal "support group" on organizational and voting matters. See *Nezavisimaya gazeta*, May 15, 2001; Ibid., May 17, 2001; and *Izvestiya*, June 16, 2001; and Robert Orttung, "With New Speaker, Putin Tightens Grip on Upper Chamber," East-West Institute, *Russian Regional Report* 6 (December 5, 2001).

[6] *Rossiiskaya gazeta*, September 5, 2000.

The reforms seemed to garner mixed results by the close of Putin's first term. On the one hand, the institutional overhaul limited the formal avenues for regional intervention into national policymaking. Except for the occasional failure to gain approval for specific legislation—such as laws related to the demarcation of land, excise taxes on liquor, and term-limits for governors—the administration was successful at using the new structures to handcuff regional leaders from blocking or overriding its initiatives. On key issues where the new Federation Council initially rejected executive branch proposals, it eventually reversed course to support the government.[7] Subverted by the predominant corporate and Moscow orientations of many new representatives of the Federation Council and with a docile Duma (the lower house of parliament), regional leaders lacked formal legal trump cards to challenge federal assaults on their *de facto* jurisdiction or to bludgeon the Kremlin into accepting their preferences on national legislation. In the eyes of the former speaker of the Federation Council, Yegor Stroev, Putin's reforms undermined the body's legitimacy and converted it into a "manageable" organ, neither responsive nor accountable to regional constituencies.[8] At most, strong provincial leaders could retain influence via "soft subversion" of federal policies at the local level; weak regional leaders, especially those heavily reliant on federal subsidies, could do little more than defer to the Kremlin's initiatives.[9]

By the same token, the State Council developed detailed sets of recommendations on several important issues—including land reform, delimitation of powers between the federal and the regional levels, defense conversion, and energy reform—that were left stillborn. None of these proposals made it out of designated working groups or were passed on for implementation by federal bodies during the organ's inaugural year. For example, proposals on the delimitation of powers that were developed under the guidance of the president of Tatarstan, Mintimir Shaimiev, were expressly rejected by the presidential administration for contributing to "the weakening of the already insufficiently effective mechanism of regulating social relations in the Russian Federation, destruction of the integrity of

---

[7] The Federation Council, for example, rejected in its first reading the new law on police that stripped regional leaders of the powers to appoint regional police chiefs. However, it eventually succumbed to executive branch pressure after only receiving the right to consult the government on the nominations submitted for presidential approval. *Kommersant*, July 21, 2001.

[8] *Novaya gazeta*, November 5, 2001. The composition of the Federation Council has been both more varied and susceptible to "outside" influence under the Putin reform. As much as 40 percent of the body consists of individuals from Moscow with only indirect associations to their "regional constituency." Furthermore, some accounts suggest that approximately 20 percent of the membership hails from big business. For more on the changing composition and increasing vulnerability of the new membership of the Council to "extra-regional" influence, see especially discussion in Robert Orrtung and Peter Reddaway, "What Do the Okrug Reforms Add Up To? Some Conclusions," unpublished draft, May 2003.

[9] The term "soft subversion" comes from Matthew Hyde, "Putin's Federal reforms and their Implications for Presidential Power in Russia," *Europe-Asia Studies* 53 (2001), pp. 729–731.

the legal system, (and) manifestations of separatism....”[10] Ultimately, the failure to realize alternative policies initiated in the State Council led many of its members to question openly the body’s legal status and substantive purpose.[11]

Despite these conspicuous setbacks, the regions were not counted out of the federal policymaking process. In principle, newly streamlined federal institutions vested regional administrations with both formal and informal channels to influence the formulation of federal policies. They ostensibly provided regional representatives with official fora to voice opinions and, in principle, to amend initiatives advanced by government ministries and agencies. In the reformed Federation Council, delegates were subject to recall by the regional executive and legislative organs that appointed them. Accordingly, there were strong political incentives for the representatives to uphold the interests of the governors or regional constituencies in reviewing federal legislation, despite their close links to Moscow.[12] As professional delegates, the new Federation Council members provided regional leaders with full-time representation in Moscow, something that they lacked in the previous body. Although they did not enjoy the national stature of their predecessors, there was nonetheless pressure on the new members to act both as lobbyists for their regions and as political “facilitators,” responsible for ironing out differences between the governors and the presidential administration.[13]

Similarly, the State Council, although a weak substitute for *ex officio* membership to the upper house, provided an official venue for the governors to meet on a regular basis. By leading working groups and standing committees, the delegates were able to spearhead the drafting and discussion of alternative reform

---

[10] Similarly, the alternative reform strategy for Russia advanced by Khabarovsk Governor Viktor Ishaev in the state Council, was stripped of the critical elements that distinguished it from the government’s program. Luydmila Romanova, “Gossovet Teryaet Politicheskii Ves,” *Nezavisimaya gazeta*, March 21, 2001.

[11] Pavel Isaev, “The State Council One Year Later,” *Russian Regional Report* 6 (September 17, 2001).

[12] Matthew Hyde, “Putin’s Federal reforms and their Implications for Presidential Power in Russia.” Newcomers to the Federation Council, as appointees of the governors and regional assemblies are demonstrating early signs of diluting the influence of the *Federatsiya* group. *Nezavisimaya gazeta*, August 7, 2001. Such constraints have lead Russian parliamentarians to advocate additional modifications to the Federation Council. For a discussion of alternative proposals that range from holding direct elections for Senators to dissolving the Federation Council altogether, see especially *Expert*, June 25, 2001.

[13] Author’s interview with several new members of the Federation Council, November 2001, Washington, D.C. In contrast, others argue that the very remoteness of the new membership weakens regional representation in the new upper chamber. Stroev, for example, asserts that because most of the new appointees are neither from the regions that they represent nor travel regularly to them, and that they tend to be Moscow insiders and foisted upon regional leaders by the “strong recommendation” of the Kremlin, the new members are conspicuously divorced from the regions. “Today, the new senators openly and fiercely state that they do not depend on the region, and demand that a law be passed that would relieve them fully of any such dependency.” *Novaya gazeta*, November 5, 2001.

programs and policies in response to executive branch initiatives, as well as to claim victories on select issues, such as the restructuring of Russia's defense-industrial complex. For example, after a joint meeting of the Security Council and the State Council, Deputy Prime Minister, Ilya Klebanov, announced that regions would receive stock in the newly formed defense corporations. He emphasized that the purpose of this decision was to ensure that regions maintain a vote in the companies that operate on their territories and that they retain an appreciable tax base.[14] In this case, the government apparently deemed it necessary to co-opt regional support by conceding a local stake in defense restructuring at the outset.

After three years of implementation, the effectiveness of Putin's federal campaign to reassert the vertical dimension of control over Russia's regions remained an open question. Despite attempts to refine presidential dominance and curb regional intervention at the federal level, the governors retained, and possibly enhanced their otherwise weak formal standing in national policymaking. What became more pronounced, however, was their reliance on indirect channels for converting this representation into influence over federal policies.

Regionalism and Russia's Nuclear Energy Policies

Consistent with the general trends outlined above, Putin's emphasis on strengthening vertical discipline and curbing the national stature of regional politicians altered sub-federal influence over nuclear energy policies. On the one hand, federal authorities were able to transcend potential regional obstacles to advance nuclear energy reforms. The failure of regional leaders to reject proposals for amending legislation on the import of spent nuclear fuel and reforming the electrical power industry, in particular, fostered images of uniform executive branch dominance.[15] With respect to the former, regional opposition within the Federation Council seemed to crumble in the face of federal pressure, as members failed to hold a formal debate on two of the three critical amendments, as well as acquiesced ultimately to the president's final approval for enacting the new legislation. The deference was conspicuous, as it contrasted starkly with heated debates during three separate readings in the Duma, numerous threats issued by key regional leaders to veto the legislation upon its arrival in the Federation Council, and evidence that the vast majority of the Russian public opposed the government's proposed amendments. A similar appearance of "railroading" federal policies over the interests of the regions was reflected by the government's early approval of the joint RAO EES/Ministry of Trade and Economic Development program for reforming the electricity sector. Notwithstanding dissatisfaction with the management of RAO EES and the company's inefficient supply of electricity

---

[14] "Regiony Poluchat v Upravlenie Pakety Oboronnykh Predpriyatii," Strana.ru website, October 30, 2001, <http://www.strana.ru>.

[15] Charles Zeigler and Henry B. Lyon, "The Politics of Nuclear Waste in Russia," *Problems of Post-Communism* 49 (July/August 2002), pp. 33–42.

across the country, Putin opted against immediately breaking up the federal grid operator along the lines recommended by an alternative report drafted by a coalition of governors under the auspices of the State Council.

Nonetheless, there were prominent regional dimensions to the new nuclear energy legislation. Despite the formal acquiescence of the Federation Council, regional interests succeeded in winning concessions to Minatom's plan for importing thousands of tons of foreign-origin spent nuclear fuel. In an effort to soften regional opposition, Minatom assumed obligations to remit 25 percent of its after-cost profits to those regions that will store the fuel, and to spend the remaining 75 percent on environmental remediation in those territories that were severely damaged by Soviet nuclear energy practices. Similarly, the inability to block the RAO EES energy reform program did not prevent key regional leaders and their Minatom allies from exploiting procedural powers in the State Council to wrangle a compromise from Putin. Much to the chagrin of the RAO EES management, Putin personally accommodated regional concerns by specifying a schedule for liberating regional operators from the predatory national electricity grid, instead of postponing the decision indefinitely.[16] Moreover, the substance of the energy reform deviated sharply from the preferences of RAO EES, which was ostensibly the most powerful player in the "game." Whereas preliminary decisions taken by the government in 2000 jibed perfectly with these interests, by the end of the process in 2003, RAO EES was forced to bow to the pressure of other actors, including Minatom, regional administrators, and commercial groups. If regional interests were fundamentally disadvantaged by Putin's federal reforms, as suggested by the marginal formal activity exercised within these newly re-constituted institutions, what explains the persistence of regional influence over Russia's nuclear energy policies?

This paradox can be explicated by distinguishing regional action from influence. As noted by others, the level of political influence over policy outcomes is not always correlated to the level of overt activity or voting behavior that takes place within an institution. In practice, "the most powerful actors, meaning those who exercise the most influence over outcomes, may be those who need to take the fewest actions."[17] This is because there are informal instruments that political actors can use to influence policy outcomes that are more subtle and indirect than a veto. Actors may exert influence by being proactive, as opposed to reactive, anticipating the probability of policy conflicts and the attendant political costs. This leaves open opportunities for regional actors to shape policymaking via a range of activities from preempting federal initiatives to raising the burden of federal accountability for subsequent policy failures. Accordingly, overt institutional activity is an inadequate indicator of regional influence, as it does not capture the informal avenues available for inserting regional preferences into

---

[16] See, for example, Illarionov's assessment of the Putin's compromise decision in *Kommersant*, July 17, 2001.

[17] See especially Lisa L. Martin, *Democratic Commitments: Legislatures and International Cooperation* (Princeton: Princeton University Press, 2000).

federal policies before an institutional crisis comes to a head. To appreciate fully the impact of regional influence on Russia's nuclear energy policies, therefore, we must examine how key players leveraged specific institutional authorities to build respective interests indirectly into the policymaking process.

Preemptive Authority and the Import of Spent Nuclear Fuel

One avenue of regional influence is strictly proactive. Instead of attempting to veto federal proposals, regional leaders can exert influence by affecting anticipated reactions.[18] Knowing that their policy proposals are subject to review by regional representatives within federal institutions, executive branch agencies have an incentive to accommodate or co-opt key regional interests in order to avoid an acrimonious, public showdown.[19] Accordingly, regional leaders can brandish new authorities within the Federation Council to exercise "preemptive authority"—via threats of raising embarrassing issues and airing public concerns—to influence indirectly the behavior of executive branch organs, such as Minatom, that are charged with drafting policies and submitting proposals for legislative approval. The key issue here is that regional influence stems less from the actual rejection of executive branch policies by the Federation Council, than from the threat of doing so. The anticipated costs of confrontation with regionally based lobbies can encourage federal agencies to take local interests into account before submitting policies for legislative deliberation and approval. By openly communicating their preferences on specific issues, regional leaders can deter federal authorities from advancing certain proposals and, in some cases, can wrangle specific commitments from government agencies to a greater extent than is otherwise reflected by their formal voting behavior in the reconstituted Federation Council.

The process of amending Russia's legislation on the import of spent fuel illustrates the dynamics of "preemptive" regional influence over Minatom policies. According to calculations by Minatom, Russia could gain over $20 billion by store and eventually reprocessing foreign-origin spent fuel. This program would be attractive to a number of European and Asian countries with nuclear power plants, since it would relieve them of the burden of storing their own radioactive waste and the costs of meeting their domestic—and likely far more restrictive—environmental regulations. Testing this market, however, required that the Russian government amend the standing legislation that banned the import of foreign-

---

[18] Ibid., pp. 9-10.

[19] John Brehm and Scott Gates, *Working, Shirking, and Sabotage: Bureaucratic Response to a Democratic Public* (Ann Arbor, MI: The University of Michigan Press, 1997), pp. 173–189. The influence of regional actors on Minatom is best captured by Brehm and Gates' metaphor of "touching the smoke detector." Here regional leaders can persuade the ministry to take their interests into account by threatening to publicly air their grievances. The ministry has an incentive to remedy the problem before it escalates into a more difficult problem, responding much like homeowners do to "smoke alarms" that notify them of a potential danger before it is necessary to call the fire department.

origin spent fuel enacted earlier to protect Russian territory from further radioactive contamination. Eager to seize on a potentially lucrative opportunity, Minatom took the lead in drafting the government's plan. The initial proposal envisaged importing up to 20,500 metric tons of spent nuclear fuel during 2001-10, which could then be stored for at least 20 years. Of this amount, 4,500 tons would be slated for final disposal in Russia while the remaining 16,000 tons would be reprocessed during 2020-40. Total expenses incurred by Minatom for the management, reprocessing, reuse, and final disposition of the spent nuclear fuel for the 40-year period were estimated at $10.5 billion. This covered $7.9 billion in operating expenses for the program, with the remainder used for investments in Minatom's research, development, and production of supplemental storage and reprocessing facilities. In addition, the initial proposal called for the projected $10.5 billion profit to be allocated liberally among federal and regional budgets. Without specifically earmarking funds, Minatom's initial proposal suggested that up to $3.3 billion could be collected via direct taxes and payments to federal and regional budgets, and that approximately $7.2 billion could be made available to redress "federal and regional socio-economic and environmental problems."[20]

At first, talk of amending Russia's legislation on importing spent fuel received mixed reviews from the regions. On the one hand, those regions that possessed nuclear power stations on their territory (all of the members of the Union of Nuclear Energy Territories and Enterprises) or that planned to construct nuclear reactors in the future generally endorsed the federal proposal. It was widely hoped that profits from the deal would translate directly into greater investments by Minatom in these regions.[21] On the other hand, approximately 30 regions threatened to reject outright Minatom's proposed amendment to the environmental legislation. Most transit regions and those provinces that were not slated to be direct beneficiaries of nuclear energy came out against the proposals. Transit regions posed a special problem in this regard, as they had legal authority to create supervisory bodies to oversee the shipment of fissile materials across their territory.[22] For example, the governor and regional legislature of Krasnodar initially sought to prohibit the transit of spent fuel across their region on grounds that the ministry's program was poorly conceived and posed a dangerous risk to public health.[23] Although most of these regions adamantly opposed the

[20] Ministry of the Russian Federation for Atomic Energy, "On the Proposed Amendment to Article 50 of the Law of the RSFSR 'On Environmental Protection'." *Technical-Economic Basis for the Law of the Russian Federation*, Moscow, 1999.

[21] *Strana.ru*, June 8, 2001. Regions with nuclear reactors included: Saratov, Sverdlovsk, Chukhotka, Tver, Murmansk, Kursk, Leningrad, Voronezh (under construction), Smolensk, Rostov (under construction), Kostroma (under construction), and Bashkortostan (under construction).

[22] *Nezavisimaya gazeta*, June 19, 2001. Prospective transit regions included: Belgorod, Irkutsk, Krasnodar, Nenetsk, Tatarstan, Rostov, Saratov, Sverdlovsk, Krasnoyarsk, Tula, and Volgodansk.

[23] *Vremya Novostei*, July 5, 2001; *Nezavisimaya gazeta,* February 20, 2001; and *Novaya gazeta*, June 25, 2001. The Governor of Novgorod, for example, stated publicly on

amendments on ecological grounds, some, such as Kemerovo, also protested out of commercial interests in protecting local industries (coal) that competed directly against nuclear energy.

In addition, the leaderships of key regions designated by Minatom as potential loci for the storage and reprocessing of spent fuel initially adopted ambivalent postures. The Chelyabinsk regional legislature, which was favorably predisposed towards Minatom's proposal, went as far as to ban a proposed local referendum on the import of spent fuel which was widely expected to come out against the proposal on ecological grounds.[24] Yet, the legislature of Sverdlovsk, an oblast with a heavy concentration of nuclear facilities, petitioned the Constitutional Court to rule on the "legality" of Minatom's proposed amendments.[25] Similarly, then-Governor of Krasnoyarsk Lebed, conditioned support for the amendment on Minatom's promise to fund construction of the RT-2 reprocessing facility and other elements of the nuclear infrastructure in his region. In doing so, however, Lebed demanded reassurance that spent nuclear fuel revenues would not be squandered by federal bureaucrats and that they would be disbursed directly to the regions that bore the brunt of the burden managing the program.[26]

Faced with potentially strong resistance in the Federation Council, Minatom deftly modified the original strategy to take into account the early variation in regional responses and to co-opt specific support. Procedurally, Minatom spearheaded the submission of a package of three bills related to the import of spent nuclear fuel to the Russian parliament. The first stipulated an amendment to the existing environmental law to allow Russia to import spent nuclear fuel; the second revised an existing law to allow Russia to lease fuel and take back the waste; and the third called for revenues generated by the program to be allocated for environmental cleanup and social programs. With respect to the last new bill, the ministry revised the substance of its earlier proposal by explicitly obligating significant funds for the environmental rehabilitation of 41 regions that had suffered severely from the legacy of the Soviet nuclear industry. According to Minatom's revised figures, those regions would receive priority allocations to include \$3.3 billion from federal and regional tax revenues, \$3.6 billion from additional payments to regional budgets, \$2.6 billion for the modernization of nuclear fuel facilities, and \$3.8 billion for environmental remediation programs.[27] In addition, regions with nuclear power stations or nuclear fuel production facilities on their territory were wooed by promises of additional investments. In a ploy to wrangle maximum benefits to the region, parliamentary and administration

numerous occasions that he categorically opposed Minatom's initial proposal and that he would not vote for the original bill if considered by the Federation Council. Interfax, April 21, 2001.

[24] *Moskovskiy komsomolets*, June 13, 2001.

[25] Newsletter of the Agency of Information Cooperation (publication of the Antinuclear Group of the Urals Ecological Union), February 2001.

[26] Interfax, June 6, 2001.

[27] *Kommersant-Daily*, December 21, 2000; and RBC News, June 7, 2001.

leaders in Chelyabinsk, for example, threatened to cave to public protests and to prohibit shipments of foreign-origin spent nuclear fuel unless the ministry formally committed to obligating proceeds directly to the regional budget.[28]

In addition, Minatom launched a concerted campaign to promote its revised proposals at the regional level. At both the ministry's initiative and behest of key regional supporters, Minatom officials conducted media blitzes and public appearances that were designed to spell out the local benefits of the spent nuclear fuel business and to "break the negative opinion" of its critics. The ministry was instrumental in the forming the Ecological Forum, a nongovernmental movement expressly devoted to hyping the benefits of the program for funding environmental remediation and to aiding the nuclear industry in providing a cost-effective alternative source of power for domestic consumption. In addition, the minister himself traveled to Krasnoyarsk both before and after the Federation Council vote to present the revised amendments at a special session of the regional legislature, and to clarify the specific mechanism for distributing projected revenues to the region.[29] Similarly, there were side deals cut with pro-Minatom regions to discount energy prices and increase the ministry's investment in local social programs.[30]

These gestures marked a significant departure from Minatom's initial plan to concentrate the proceeds from the spent nuclear fuel business primarily in the federal government's hands. By promising to distribute "pork" prior to the Federation Council vote, Minatom was able to split the provincial votes enough to ensure that the opposition could not block the passage of the laws. Initially, the Federation Council preferred not to act on the proposed amendments, as regional opponents doubted they could win, while proponents of the legislation were loathe

---

[28] Although regional officials initially demanded that 30 percent of the profits be distributed to the region, they settled for a formal commitment to allocate 25 percent to cover the region's social and environmental programs. In addition, the prospects for extracting greater federal investment in the regional energy infrastructure seemed to be one of the reasons why Saratov Governor, Dmitri Ayatskov, the chairman of the Union of Nuclear Energy Territories and Enterprises, actively supported the proposed amendments. Interfax, June 7, 2001; "Vvoz yadernogo Topliva: Minatom Ishchet Podderzhki u Gubernatorov," Strana.ru website, June 7, 2001, <http://www.strana.ru>.

[29] *Moskovskiy komsomolets*, June 13, 2001. The minister's visit to Krasnoyarsk krai was especially important, given that the speaker of the regional assembly sharply criticized Minatom's proposed amendments and voted against them in the Federation Council. The fact that the minister returned to the region shortly after the vote was interpreted widely as reflecting the Minatom leadership's respect for the speaker and its lingering concerns about building a constituency in the regional legislature to win over his support. *Krasnoyarskiy rabochiy*, June 6, 2001; and *Nezavisimaya gazeta*, August 3, 2001.

[30] Such regions included Saratov, Sverdlovsk, Tomsk, Chelyabinsk, Voronezh, and Nizhniy Novgorod. See discussion in Ann MacLachlan, "Russian Government Calls on Regions to Weigh in on Spent Nuclear Fuel Import Project," *NuclearFuel* 25 (April 17, 2000), <http://www.mhenry.com>; and Igor Kudrik, "Russian Regions to Decide on Fuel Imports," Bellona Foundation website, April 11, 2001, <http://www.bellona.no>. "Vvoz yadernogo Topliva: Minatom Ishchet Podderzhki u Gubernatorov," Strana.ru website, June 7, 2001, <http://www.strana.ru>.

to make their support public, given that polls at the time revealed that over 90 percent of the Russian population did not support the government's proposed amendments. Ultimately, however, the upper chamber could not avoid the issue altogether, and was obliged to vote on the least controversial environmental remediation bill. The legislative package was then forwarded directly to President Putin, who approved it.[31]

The politics surrounding the spent fuel amendments suggest that the vote did not rebuff regional interests altogether. While it marked a political defeat for those opposition regions that stood to receive little or nothing from the proposed deal, it constituted a victory for other provinces that hoped to benefit both directly and indirectly from the amended legislation. Similarly, the peculiar decision by the Federation Council to endorse one bill and pass the other two directly onto Putin did not represent an institutional defeat for the body vis-à-vis the government. Given the previous problems that the legislation encountered in the Duma and in the court of public opinion, Minatom almost certainly risked escalating an acrimonious political debate without infusing its proposals with tactical inducements to key regions. By preempting a regional backlash and averting a public showdown in the Federation Council, it was able to secure support for its preferred legislation, albeit with concessions to critical regional interests. This not only strengthened Minatom's hand for entering the global spent nuclear fuel market and boosting investment in domestic nuclear power generation, but helped to ensure that the resulting revenues would be allocated specifically for the remediation of regional environmental and economic legacies of the Soviet nuclear program. It also allowed regional leaders to take credit for these prospective local payoffs, while simultaneously shifting blame onto the federal government, and potentially onto Putin personally, for any future public backlash, broken promises, or adverse consequences linked to the import, transit, storage, and reprocessing of spent nuclear fuel.[32]

Coalition Building

Putin's federal reforms also created opportunities for the regions to affect national policies by forging ad-hoc political coalitions with socio-economic interest groups at both the regional and federal levels. With reform of the Russian energy monopoly, RAO EES, which controlled a large portion of power generation and all of the power transmission and distribution networks, regional leaders found ways to circumvent the strictures of the new federal system and to weigh in on the national debate.

On its face, the battle over electricity reform was resolved at the federal level without discernable input from the regions (Putin rejected the alternative energy plan advanced within the State Council). In practice, however, the debate was neither a strictly "center versus regions" issue, nor was it settled without regard to

[31] *Kommersant-Daily*, June 30, 2001.
[32] *Nezavisimaya gazeta*, June 22, 2001.

key regional interests. Regional administrations were able to affect the outcome of the debate by aligning themselves with Minatom and its energy-producing agency, Rosenergoatom.[33] The coalition was premised on shared interests related to cheaper energy supplies, freedom from the overbearing influence of RAO EES, generation of commercial and employment opportunities, and opposition to environmentalists. Although the State Council served as a weak instrument for formally representing regional interests, it proved an effective forum for forging new coalitions and lobbying the government that complemented other avenues of influence established through the national parliament and control over local media outlets. Taken together, these institutional mechanisms amplified the political salience of regional concerns, empowering actors from across the whole of Russia to recognize and combine mutual interests to support or challenge federal policy initiatives.

The debate over Russia's electricity reform unfolded against the backdrop of a deepening national energy crisis in 1999–2000 and mounting frustration with the existing transmission and pricing mechanisms for power generation. The crisis was precipitated by the industrial recovery that followed the August 1998 financial meltdown, and the attendant jump in prices for foreign goods that stimulated domestic production. While there was general agreement that energy generation and transmission had to be radically changed, there was no consensus on the ends or means of electricity reform. The debate was sharpened by the attempts of RAO EES, which at the time controlled 75 percent of Russia's energy infrastructure and covered approximately 80 percent of the national electric power market, to exploit the improved economic situation to realize its two long-standing objectives. First, it sought to increase wholesale prices or at least to sustain them at a sufficient level to generate much-needed revenue for the refurbishment and expansion of the country's power-generation capacity. Second, RAO EES moved to secure repayment of outstanding debts incurred by regional intermediaries that were responsible for directly distributing energy and collecting payment from local customers.[34] Due to the lack of efficient court/arbitration and bankruptcy

---

[33] Minatom's energy generation arm, Rosenergoatom, can be viewed as a corporate actor, as only five percent of its investment in NPPs came from the federal budget. The rest was generated by Minatom itself, both from selling power and from other operations (such as the HEU deal with the United States). See Interview with Yevgeni Adamov, RBC News, November 22, 2000.

[34] RAO EES sold energy to regional companies that distributed it to individual, government, or corporate customers and collected payments. It is well known that customers often did not pay. This was especially widespread among military facilities and unprofitable state-owned enterprises. But even when customers paid (individual customers were considered as having the best record among others), regional companies often simply pocketed the money and did not transfer it to RAO EES. As a result, non-payments were simply abysmal from the national monopoly's perspective. Attempts to reign in regional companies were usually fruitless because in many cases the regional distributors were protected by local governments. An extreme case of such behavior purportedly involved the Amur Oblast division of RAO EES, whose successive chiefs siphoned money from the company and the

procedures, local customers (both enterprises and individuals) had to bear the brunt of RAO EES' assault on the intermediaries to collect on these debts. This usually involved switching off power to chronically delinquent regions.

In several cases the nuclear infrastructure, both civilian and defense-related, was directly affected by shutdowns, which were supposed to force delinquent customers or regional authorities into payment. For example, RAO EES confounded operations at the Dimitrovgrad nuclear reactor in Ulyanovsk Oblast by cutting off the power supply, risking a nuclear meltdown. Local authorities intervened, however, to resolve the crisis and avert an accident.[35] Similarly, RAO EES reduced the power supply to the Sibvolokno chemical plant, located in the nuclear closed city, Krasnoyarsk-45, an action that nearly led to the release of toxic gases into the atmosphere.[36] Even as late as in 2002, power was switched off at a chemical plant, Tverkhimvolokno, that produces components for solid-fuel rocket motors. Although Putin personally intervened to order the restoration of power to the facility, the crisis nearly cost Russia the ability to produce solid-fuel missiles, including the mainstay of its strategic forces, Topol-M ICBM.[37] Regional administrations attempted to counter with demands for compensation from RAO EES affiliates for the construction and maintenance of local power lines to offset debts, but this tactic was not efficient, especially since each regional or local government had to face the federal-level monopoly on its own.[38]

Minatom, which accounted for about 15 percent of the country's energy supply, had its own grievances with the energy monopoly.[39] Since nuclear power plants (NPPs) were forced to sell energy through the RAO EES-controlled Federal Wholesale Market of Electric Energy and Power (FOREM), their commercial activities effectively fell under control of RAO EES in several important respects. First, the NPPs were deprived of the right to set prices for the electricity that they

---

region, and sold the Ministry of Defense's debt to a private company for an enormous discount. RAO EES was unable to crackdown on them despite many complaints and criminal investigations conducted by local authorities. *Izvestiya*, October 19, 2000.

[35] Ibid.; and RBC News, November 4, 2000. There is no evidence, however, that prevention of a nuclear accident was the main concern of the local authorities.

[36] The situation was corrected with the help of Yevgeni Adamov, who called Krasnoyarsk Governor Alexander Lebed and explained to him the risks for the city and oblast, see *Izvestiya*, February 17, 2001; and ITAR-TASS, February 15, 2001. It seemed significant that plant management decided to involve Adamov, even though the plant itself does not belong to the Minatom system.

[37] *Izvestiya*, December 22, 2003.

[38] One such case was the city of Kurgan. See *Izvestiya*, November 23, 2000.

[39] In 2001, NPPs covered 15 percent of the wholesale electric power market, while RAO EES supplied more than 80 percent. Rashid Alimov, "Unified Electrical Systems (UESR): 'It is Not our Duty to Support the Nuclear Energy Industry'," Bellona Foundation, September 28, 2001. The remaining five or less percent was covered by small-scale local power producers. The distribution did not change much by 2003: 16 percent for Minatom and 84 percent for RAO EES. Mark Kramer, "Restructuring of the Russian Electricity Industry," PONARS Policy Memo No. 304, November 2003.

produced. RAO EES sold energy purchased from nuclear power stations at the same price as electricity that was generated by thermal power stations, even though energy produced by NPPs was considerably cheaper. The difference was masked by accounting "tricks." For example, the price of NPP-generated energy included the operating and investment costs of Rosenergoatom, whereas for RAO EES-operated power stations the same component was levied by its regional operators separately and above the price of energy.[40] This effectively deprived Minatom from exercising its competitive advantage to expand its share of the market and potentially amassing enough profits to complete or construct new NPPs. As late as 2002, Rosenergoatom complained that control of power distribution and prices by RAO EES deprived Minatom of considerable revenues: whereas investment needs for 2003 amounted to 30.7 billion rubles, in 2002 the concern generated only 22.2 billion that could be invested.[41]

Second, NPPs were administratively divorced from their local customers. In practice, RAO EES was able to manipulate the assignment of payees through FOREM, which pooled not only energy, but also payments to NPPs. As a result, Rosenergoatom was paid not by the actual customers of its energy, but by other entities. Accordingly, Rosenergoatom consistently complained that it was assigned the worst-paying customers and often those that did not pay cash.[42] Cash-paying customers were routinely assigned to power plants controlled by RAO EES—a vital benefit in an economy, where cash was scarce and often replaced by other, non-monetary forms of payment. In January 2000, for example, NPPs received only 14 percent in cash payments. By the end of the year, as the health of the Russian economy began to improve, cash payments to NPPs gradually increased to cover 60 percent of the debt, but thermal and hydroelectric power stations were still ahead with 100 percent and 85–100 percent, respectively, of their cash payments.[43] This preferential treatment led to a series of lawsuits, including a case

---

[40] According to FOREM data, the cost of 1 kilowatt of NPP-generated electric power is on average 19.2 kopeks, whereas fossil fuel power stations produce 1 kilowatt at a cost of 36.6 kopeks. By comparison, the cost of oil power plants is the highest at 72.7 kopeks per 1 kilowatt. The operating and investment components for Rosenergoatom in 2001 amounted to 35.2 kopeks per kilowatt, while for fossil fuel power plants it was 40.8 kopeks per kilowatt. See *Nezavisimaya Gazeta*, August 8, 2001. The former Minister of Atomic Energy, Yevgeni Adamov, claimed in 2001 that NPPs sold power to RAO EES for 20 kopeks per kilowatt, while the latter resold it to individual customers for 62 kopeks per kilowatt and to industrial customers for 74 kopeks per kilowatt, thus pocketing the difference. See the speech of Yevgeni Adamov in Ozersk, in *Ozerskaya Panorama*, March 23, 2001.

[41] *Kommersant*, January 29, 2003.

[42] *Vedomosti*, March 1, 2001.

[43] *Nezavisimaya Gazeta*, October 31, 2000, p. 4. In the first half of 2001 the situation improved marginally, as approximately 75 percent of power sold to FOREM by the NPPs was paid in full. See *Izvestiya*, November 3, 2000; and *Kommersant,* July 4, 2001; *Kommersant*, July 13, 2001; and *Kommersant*, November 22, 2001.

in Penza where the local court ruled in favor of the Balakovo NPP, as well as a similar, but unsuccessful case in St. Petersburg.[44]

The conflict escalated to the point where Rosenergoatom threatened to sever ties with RAO EES and to sell power independently. When this plan did not materialize, Rosenergoatom attempted to withhold power supplies to RAO EES until NPPs enjoyed the same treatment as non-nuclear power plants.[45] In fall 2000, Rosenergoatom threatened to restrict deliveries by 22 percent (it eventually reduced power supplies by 45 percent).[46] In response, RAO EES reduced purchases of power from NPPs, which, in turn, forced an overall decline in nuclear power generation.[47] The conflict came to a head in summer 2001, as Rosenergoatom threatened to take RAO EES to court and announced that it would limit the supply of electricity commensurate to the cash payments that it received during the preceding week.[48] However, NPPs were at a disadvantage in this conflict because they could not credibly threaten to stop or power down nuclear reactors; their only hope for fair treatment rested with wholesale reform of the energy sector.

Independence from RAO EES was a vital precondition for Minatom's ambitious development plans. The program for developing the nuclear energy sector was adopted by the Russian government in May 2000. This envisioned a $15 billion investment in nuclear power generation over the next 10 years that, in turn, would position Minatom to raise its share of electricity production from 15 to 40 percent across the country (and from 30 to 45 percent in the European part of Russia).[49] Minatom's plans for the next decade also called for reversing the late 1980s-early 1990s freeze on reactor construction and for commissioning on average one new reactor per year. This included the completion of up to nine reactors by 2010 (most of which had been put on hold during the previous decade) and construction/replacement of more than 25 reactors during 2010-20.[50] Revenue from power generation was particularly vital since federal funding could only support the construction of three to four reactors until 2010.[51]

---

[44] In July 2000 the Leningrad NPP went to court, but without much success. See RBC News, November 11, 2000.

[45] *Kommersant*, September 27, 2000.

[46] Power production was restored after Deputy Premier Viktor Khristenko ordered RAO EES to normalize the payments situation to the NPPs. See RBC News, November 21, 2000.

[47] RBC News, December 12, 2001.

[48] "'Rosenergoatom' Gotovitsya Ogranichit Elektrosnabzhenie RAO EES," Strana.Ru, July 3, 2001.

[49] These ambitious plans were subsequently tempered, as 35 percent was considered, according to a 1991 law, "dominant" or "monopolistic." "Atomshchiki Vidyat Svoe Reshenie Energeticheskikh Problem," Strana.ru website, October 18, 2000; and RBC News, December 19, 2000.

[50] "Strategiya razvitiya atomnoi energetiki Rossii v pervoi polovine XXI veka," *Moscow: TSNIIatominform*, 2001, p. 57; *Nezavisimaya gazeta*, April 26, 2001; and RBC News, June 1, 2001.

[51] *Nezavisimaya gazeta*, December 25, 2003.

It should be noted that Minatom's strategy for aggressive expansion could have led to the same monopolistic position as that of RAO EES. Anatoli Chubais, the head of RAO EES, opined that within the wholesale energy market (i.e., excluding direct contracts with customers) Minatom accounted for 40 percent, rather than 15 percent, of the nationwide supply of electricity, and up to 70 percent of deliveries to the European part of Russia.[52] Accordingly, he argued that Minatom was well poised to corner the wholesale market in the absence of strict government regulation.[53]

At the same time, the prospects for lower prices offered by NPPs and the relative independence from RAO EES made Minatom an attractive political ally for numerous regional governments. Many regions not only supported existing nuclear power stations on their territories but lobbied heavily for the construction of new reactors to meet the growing demand for energy. Alignment with Minatom was all the more attractive because roughly 10 percent of the construction costs for a nuclear reactor covered the social infrastructure of the host region.[54] Then-Minister of Atomic Energy Evgeniy Adamov proudly boasted that requests for new NPPs and/or reactors came from across the country, including Tatarstan, Bashkortostan, Archangelsk, and the southern Urals. Bashkortostan President Midkhat Rakhimov claimed that by the republic's own calculations it could not cover more than half of the projected demand; only an NPP could possibly provide sufficient energy.[55] In Arkhangelsk, the governor's office decided to build a nuclear heating plant to supply Severodvinsk, the home of one of the largest shipyards in Russia, that effectively reversed the 1990 decision by the Arkhangelsk city council to create a nuclear-free zone on the territory.[56] Similarly, the leaders of Sverdlovsk signed a special agreement with Minatom to construct a fourth reactor at the Beloyarsk NPP.[57] Moreover, in some cases, Minatom was able to forge alliances with regional administrations to suppress local environmental groups. This endeavor was especially successful in Kostroma, where Minatom officials worked closely with regional leaders to overturn the ruling of the 1995 local referendum that banned construction of a nuclear power station on the territory.[58]

---

[52] RBC News, December 15, 2000. By law, Minatom has been permitted to sell wholesale power via FOREM.

[53] The 40 percent figure mentioned by Chubais was probably correct because it was exactly the share of TsDR that Adamov demanded be given to Minatom. Ironically, the new minister of atomic energy, Alexander Rumyantsev, indirectly confirmed the accusations of monopolism, noting that the share of 30 percent in the power supply in a region should be sufficient to warrant attention of anti-monopoly regulators. Interview with Alexander Rumyantsev in *Vek*, May 25, 2001, p. 9. As noted above, in some regions this share is already about 30 percent and is expected to increase with new construction.

[54] The figure was quoted by the governor of Saratov oblast, Dmitri Ayatskov. In *Nezavisimaya gazeta*, December 25, 2003.

[55] *Nezavisimaya gazeta*, December 15, 2001, p. 4.

[56] ITAR-TASS, June 6, 2001.

[57] Interview with Sverdlovsk Oblast Governor, Edward Rossel, cited in *Vek*, July 13, 2001.

[58] *Itogi*, April 2, 2001.

A particularly close political bond was established between Minatom and the regional administration in Tomsk, which faced the closure of two plutonium-producing nuclear reactors at the Siberian Chemical Combine (SKhK) in Seversk (former Tomsk-7). Under this measure, which was strongly advocated by both the Russian federal government and by the United States, the oblast stood to lose approximately 30 percent of its electricity supply. From the outset, the regional administration was determined to retard implementation of the plan. In 1998, Tomsk Governor Viktor Kress succeeded at convincing the Minister of Atomic Energy to keep the reactors running, in return for the promise to lobby GAN to restore the full capacity of one of the reactors that was earlier damaged in an accident.[59] In a conspicuous gesture of cooperation between the region and Minatom, Governor Kress sent a letter to then-Prime Minister Sergei Stepashin, demanding that GAN drop its objections so that the reactor could be reopened.

After a 2000 U.S.-Russian agreement stipulated the inevitable shutdown of these reactors, the Tomsk leadership continued to work closely with Minatom to lobby for an alternative electricity provider for the region. It advocated construction of a new nuclear heating station (AST-500), previously authorized by Minatom in 1996, over a new coal-burning power plant that could have rendered the oblast dependent upon Tomskenergo, the regional division of RAO EES. Since the AST-500 was a small-scale project that was originally slated to be funded directly from the local budget, the oblast administration was open to exploring the prospects for developing a full-scale NPP that could attract outside funding and provide the basis for cultivating a closer relationship with Minatom. Ultimately, however, plans for both an NPP and the AST-500 were scrapped when Moscow succeeded at reaching an agreement with the United States in March 2003 to provide funds for the construction of a fossil-fuel power plant to replace the plutonium-producing reactor.[60]

In addition, Minatom sought to increase its influence over the wholesale electricity market by reducing RAO EES' 80 percent control over the single accounts center (Tsentr Dogovorov i Raschetov, or TsDR), the body charged with regulating cash payments across the electricity market. In particular, Minatom demanded to increase its share from 20 to 40 percent. It also threatened to sell cheap energy directly to customers, thus allowing the nuclear industry to break into RAO EES's monopoly and undermine efforts at creating a controlled market.[61] In this spirit, Minatom decided to form a single power generation company that would control all nuclear power stations in Russia and boost investment in the construction of additional reactors by optimizing existing and expected cash

---

[59] *Nezavisimaya gazeta*, October 3, 2000.

[60] *Ibid*; See also Sergei Novokshenov, "Skinemsya Vsem Mirom na Novuyu AES," February 21, 2001, <http://www.minatom.ru>. This article appeared in a local Tomsk newspaper.

[61] *Vedomosti*, March 1, 2001.

flows.[62] The plan was implemented in early 2002, when Rosenergoatom was considerably enlarged to incorporate one of the Leningrad NPPs (which was previously managed directly by Minatom), as well as enterprises and institutes that provided technical and research support for NPPs. In addition, all future new reactors were slated to fall under the administrative auspices of Rosenergoatom.[63]

The first stage of reform dealt with the issue of preserving the national power grid. Minatom's initial plan for energy reform, introduced in fall 2000, called for nationalizing the country's energy grid and allowing NPPs to sell power directly to customers.[64] This scheme would have enabled Minatom to set its own pricing policy and collect payments directly, thus undercutting RAO EES's monopoly. In contrast, the RAO EES plan envisioned formation of a national power transmission company with 75 percent plus one share controlled by the state. This proposal was supported by Minister of Economy and Foreign Trade German Gref, the leading liberal economist in the Putin government, who rejected Minatom's program on the grounds that the government could not afford to nationalize the electricity sector.[65] With this political support, RAO EES won the first battle, as a special interagency meeting in December 2000 streamlined RAO EES and ceded to the government 75 plus one percent shares in the transmission company.[66]

At that stage of the game, a special group in the State Council chaired by Tomsk Governor Kress, prepared its own version for reforms. This group recommended the creation of 30–40 "vertically integrated" companies that would control power generation and distribution within each area consisting of two or three regions.[67] Although the plan stipulated preservation of a wholesale market and central pricing authority, the new regional companies were slated to fall under the jurisdiction of the governors. This push for decentralizing control over electrical power generation and transmission was very popular among the governors. For example, Khabarovsk Kray Governor Viktor Ishaev proposed that Putin turn the regional division of RAO EES directly over to regional administrations (which Putin refused).[68] In general, the plan contained several market-oriented elements, as it stipulated greater accountability and transparency in supplying electricity, the elimination of "cross-accounting" (the practice of

---

[62] According to Minatom's press service, investment in power generation was distributed in the following manner: 40 percent comes from Rosenergoatom's direct energy sales, 40 percent comes from Minatom's "other sources," and only five percent from the state budget. Press Service of Minatom, April 16, 2001, <http://www.minatom.ru>.

[63] The official name is Federalnoe gosudarstvennoe unitarnoe predpriyatie kkontsern Rosenergoatom (the Federal State Unitary Enterprise, Rosenergoatom Concern).

[64] *Kommersant*, September 27, 2000; and RBC News, September 29, 2000.

[65] *Kommersant*, November 22, 2000. Since RAO EES was partially owned by private, including foreign, investors, the government would have been obligated to buy their shares at market price; even worse, there is as yet no law on nationalization, so the precise procedure does not exist. See *Kommersant*, November 15, 2000.

[66] *Izvestiya*, December 8, 2000.

[67] Interview with Viktor Kress, Strana.ru website, March 1, 2001.

[68] *Izvestiya*, February 28, 2001.

mixing all energy and money payments), and the lowering of artificial barriers to market entry.[69] This plan would have preserved Minatom as a unified national-level power generation company, and would have allowed the establishment of direct contracts between Minatom and area-wide "vertically integrated" companies (and, by implication, directly with regions).

Under pressure from Minatom and the regions, Prime Minister Mikhail Kasyanov revised the initial decree in July 2001. The new document sanctioned the formation of a single nuclear power generation company, Rosenergoatom, against the strong objection of RAO EES.[70] Minatom also received the right to export electrical power without coordinating with RAO EES.[71]

RAO EES' perceived victory triggered intense political infighting within the government. Andrei Illarionov, a staunch proponent of rapid marketization among Putin's economic advisors, complained that Kasyanov's rejection of the State Council program threatened to undermine the energy sector and worsen the investment climate in Russia, as well as failed to limit the predatory control exerted by RAO EES.[72] At the same time, the "Kress Plan" seemed to gather more support from the regions. Independent analysis of the impact of the government's plan on Krasnoyarsk Kray, for example, revealed that it would lead to the closure of small fossil-fuel power plants and that an independent hydroelectric power station would fall under control of Krasnoyarskenergo, a division of RAO EES, reinforcing the concern's regional monopoly.[73] Accordingly, that region immediately produced several alternative plans. Then-Governor of Krasnoyarsk Valeri Zubov envisioned creating a joint Krasnoyarsk-Khakasia power generation company, in effect *de facto* implementing the "Kress Plan" without federal endorsement. Metallurgical enterprises in Krasnoyarsk, long-time adversaries of Chubais, began to explore the prospects for forming an independent electrical power generation company that would be able to utilize federally owned transmission assets once they were hived off into independent companies.[74]

Similarly, the presidents of Bashkortostan and Tatarstan began to explore new approaches for merging their respective energy resources that were consistent with the "Kress Plan." Moscow Mayor Yuri Luzhkov requested that Rosenergoatom serve as an energy provider to the capital city outside of RAO EES channels, offering to finance construction of a new electrical power line that would connect Moscow to a family of NPPs. Therefore, in spite of the government's ruling, the governors continued to view Minatom as a valuable ally against RAO EES, as well as an alternative source of "cheap" energy.

---

[69] Ivan Goryaev, "Restrukturnuli," Grani.ru website, May 24, 2001, <http://www.grani.ru>.
[70] *Vremya Novostei*, July 5, 2001.
[71] *Kommersant*, July 4, 2001. The first foreign customers for Rosenergoatom electricity were Georgia and Ukraine.
[72] Andrei Illarionov: "Poslednii Variant Reformy Elektroenergetiki Ne Byl Soglasovan s Gossovetom," *Strana.Ru*, June 29, 2001.
[73] *Nezavisimaya gazeta*, July 19, 2001.
[74] Ibid.

The opposition to RAO EES' proposal had far-reaching implications. The defeat of the "Kress Plan" provoked several influential regional leaders— such as Leonid Polezhaev (Omsk), Edward Rossel (Sverdlovsk), and Mintimir Shaimiev (Tatarstan)—to propose the formation of a professional staff for the State Council and mandatory meetings with vice-premiers on the issue.[75] In effect, conflict over the reform of RAO EES jeopardized Putin's plan for fundamentally revising the country's federal structure.

This alliance between Minatom and regions was subsequently tested in the political contest over the composition of the Unified Tariff Body (ETO), the agency authorized to establish electricity tariffs. Although RAO EES maneuvered in an attempt to dominate this body in the same way that it did with the TsDR, victory was not assured. In addition to Minatom, a powerful group of aluminum producers and regional leaders emerged to challenge the electricity monopoly in an effort to establish a more equitable procedure for setting energy tariffs. In August 2001, Putin signed a decree that established the ETO and tasked it with coordinating new tariffs with region administrations.[76]

The next stage of the political battle unfolded over the organization of the Administration of the Trading System (ATS)—the electricity trading center that regulated domestic buying and selling of energy at competitive rates. This process involved the active participation of not only regional administrations, but regionally-based commercial enterprises and industrial consumers of power, such as local aluminum producers. Due to their high economic and political profile, the convergent interests of these business entities and regional administrations were mutually reinforcing, which strengthened the hand of regional bidding in the reform process.

At the outset, RAO EES insisted on controlling 80 percent of the votes on the ATS Supervisory Council and denied of any role for customers. In contrast, the union of Rosenergoatom and aluminum producers fought for a larger share for themselves (Rosenergoatom demanded at least 20 percent for itself and another 20 for its allies) and for restricting RAO EES to only 25 percent of the votes. The final compromise, yet again, offered a temporary victory for RAO EES but a long-term victory for Minatom and its regional allies (represented by both regional energy commissions and corporate consumers of energy). Specifically, a candidate from RAO EES was elected as the ATS administrator and RAO EES itself was allotted 50 percent of the votes in the ATS Supervisory Council. However, by summer 2003 its share was supposed to dwindle to 35 percent with the balance to be distributed among other members, such as Rosenergoatom.[77] The subsequent fight over control of the ATS resulted in further losses for RAO EES, as its share was

---

75 Ibid., June 9, 2001.

76 "President Utverdil Kontseptsiuy Edinogo Tarifnogo Organa," Strana.ru website, August 6, 2001, <http://www.strana.ru>.

77 Sergei Pletnev, "NaRynke Elektroenergii Poyavilsya Direktor," Strana.ru website, November 23, 2001, <http://www.strana.ru>.

reduced to 35 percent on January 1, 2003, about six months earlier than originally projected.[78]

In April 2003, ATS took control of FOREM and began to prepare for the gradual transition toward a segmented market. Free bidding for contracts began on November 1, 2003, but Rosenergoatom initially refrained from participation citing uncertainty over its ability to sign contracts directly with customers due to the fact that RAO EES continued to exercise considerable control over decision-making within ATS. Rosenergoatom representatives cited, in particular, familiar concerns, including pricing and the ability to conclude direct, long-term contracts with customers.[79] These concerns were quickly put aside by December 11, when Rosenergoatom began to participate in free-market sale of energy. According to the company's spokesman, it was able to generate three times more revenue (charging lower prices) in the first days than would have been possible within the regulated sector of the market.[80]

Conclusion

Putin's initial federal reforms raised more questions than answers. On the one hand, the stature of federal offices in the regions conspicuously declined, while the ability of the governors to act at the national level formally constricted. In contrast to the lack of consensus on basic questions of center-periphery relations and the arbitrary penetration of federal policymaking by regional interests under Yeltsin, the Putin administration succeeded early on in the first terms at clarifying the political boundaries of the federal dialogue and strengthening the vertical dimension of executive power. Yet beyond these broad contours, uncertainty persisted over the direction and effectiveness of the federal reforms. Notwithstanding the shift in tone of center-periphery relations, there was little evidence that the Kremlin proved adept at stemming regional clout at the center, especially given the fluid form and substance of reconstituted federal organs.

As suggested by the revisions to federal nuclear energy policies, the regions were down but certainly not out of the national policymaking process in the wake of Putin's initial assault. The very reforms of the Federation Council and State Council secured for the governors a structural role in the decisionmaking process, affording them indirect and substantive influence over the government's nuclear

---

[78] Guzel Fazullina, "Kakim budet Optovyi Rynok Elektroenergii Stanet Yasno k Iuyluy," Strana.ru website, March 5, 2002, <http://www.strana.ru>. RAO EES' share was expected to decline even further to 25 percent, as new players joined the energy market. For a comprehensive analysis of the energy reform see especially *Kommersant,* June 4, 2003.

[79] *Izvestiya*, November 4, 2003. The plan envisioned that part of the market should remain regulated, while initially up to 15 percent should function exclusively on market principles.

[80] *Nezavisimaya gazeta*, December 22, 2003. According to the Chairman of the Board of ATS, Dmitri Ponomarev, price in the free-market sector was, on the average, 18 rubles/MW lower than in the regulated sector. See *Kommersant*, December 12, 2003.

energy policies. In particular, regional leaders moved quickly to exploit their new institutional authority to draw public attention to controversial federal policies and to set the terms for the ensuing national debates. By chairing committees assigned to investigate and review specific aspects of federal proposals, regional leaders and their delegates were well poised to set the agenda for drafting alternative nuclear energy programs. As evidenced by the process of amending legislation on the import of spent nuclear fuel, key regional leaders succeeded in wielding new authority in the federal assembly to preempt Minatom's preferred policy proposals and to wrangle federal inducements, notwithstanding the Federation Council's formal deferral to the government's revised program. As reflected in the battle over electricity reform, regional leaders also were adroit at using the new State Council as a forum to cement a political alliance with Minatom to lobby against the country's energy monopoly by advancing an alternative energy reform package. Without dictating the terms of energy reform, the regional-Minatom coalition succeeded at raising the political costs to the Putin administration of embracing the electricity monopoly's preferred platform, as well as at securing future executive branch support for breaking RAO EES's grip over the national grid system. Both episodes demonstrated that the influence of regional leaders over national legislation extended beyond formal voting behavior in the modified federal institutions.

An important consequence of this burgeoning "federal face" of regional influence was the introduction of additional transparency into Russia's nuclear energy policymaking. Putin's final approval of the spent nuclear fuel legislation, for instance, was contingent upon Minatom's future compliance with its commitments to the regions. Spent nuclear fuel could be received by Russia "only after appropriate state environmental examinations have been carried out, if other radiation risks will be lowered, and if the level of environmental security will be enhanced as a result of the implementation of such a project."[81] In effect, this obligated Minatom to disclose future regional impact statements and subject them to additional public and legislative scrutiny.

In deference to public and regional pressure, Putin also submitted to parliament a new bill that called for all spent nuclear fuel import contracts to be approved by a newly formed government commission. In signing the legislation, Putin decreed this commission to consist of five members from the presidential administration, five members from the government, five members from the Duma, and five members from the Federation Council. The organ was vested with formal authority to: 1) review the management of spent nuclear fuel imports; 2) submit to both branches of government an independent assessment of Minatom's performance; and 3) oversee the disbursement of proceeds from the contracts into a special fund to ensure that they "would not dissolve into the state budget." In addition, it was officially stipulated that spent nuclear fuel contracts would require special permits from the Customs Committee, as well as certification and a license from GAN.[82]

---

[81] Interfax, December 20, 2001.
[82] ITAR-TASS, December 20, 2001; and *Vek*, June 20, 2001.

These institutional measures provided openings for greater transparency of Minatom's actions in the spent nuclear fuel business. Emboldened by the oversight provisions in the final legislation, the head of GAN turned his agency into a close watchdog of Minatom's regional operations. Not only did he personally challenge Minatom's claims regarding Russia's technical and administrative capacities to safely manage additional spent nuclear fuel imports, but GAN worked directly with regional politicians in Murmansk, Krasnoyarsk, and Chelyabinsk to uncover violations in the assembly and transport of spent fuel across Russia.[83] These efforts resulted in the detection of repeated violations of medium- and low-level active nuclear waste discharges into the local reservoirs surrounding Minatom's flagship Mayak storage depot. In response, GAN delayed certifying the facility's renewed license for further reprocessing from the end of 2002 to March 2003. Vishnevskiy specifically cited violations of the amended Article 50 and the potential for aggravating existing problems with the import of foreign spent fuel in support of the decision.[84] Although Vishnevskiy was unceremoniously "retired" from his post in July 2003, his successor, the former head of Minatom's foreign construction branch and a proponent of the spent fuel import program, nonetheless went out of his way to endorse stringent regulatory guidelines for the agency's enforcement of each contract.[85] Similarly, environmental groups seized on these levers to rally regional protest movements over the ensuing two years.[86]

---

[83] *Izvestiya,* December 28 2001; and Charles Digges, "Minatom's Starry-Eyed Import Plans Defy Safety Imperatives and Russian Sense," Bellona Foundation website, July 25, 2002, <http://www.bellona.no>.

[84] *Trud*, January 15, 2003; and "Broad Overview of 2002 Gosatomnadzor Activities and Achievements," February 20, 2003, <http://www.gan.ru>; in *FBIS-SOV CEP 20030528000229*. Although impressed with GAN's oversight, environmentalists questioned Minatom's motives for acceding to the regulatory body's decision. First, they interpreted it as a ploy to acquire additional resources for constructing a new nuclear power plant that could "vaporize" the contaminated water in the region. Second, they suspected that Minatom intended to use GAN's ruling to woo Washington's consent for Russia's future import of American origin spent nuclear fuel. Because Washington stipulated a preference for long-term storage over reprocessing, environmentalists assumed that GAN and Minatom colluded to close Mayak's reprocessing facility in order to demonstrate to the Americans that Russia is not intent on reprocessing American-origin spent nuclear fuel. See discussion in *Novaya gazeta*, January 16, 2003. This, however, disregards: the acrimony between GAN and Minatom; Minatom's stated commitment to reprocessing; the legal prohibition on burial of nuclear waste in Russia; and the U.S. fixation on tying consent to Minatom's cessation of nuclear cooperation with Iran.

[85] For background on GAN's new management team, see especially Charles Digges, "Malyshev Makes His First Public Statement Amid Mixed reviews From His Colleagues,"Bellona Foundation website, July 24, 2003, <http://www.bellona.no>.

[86] For example, Greenpeace mounted public campaigns to expose the shipping routes that were slated to be used for importing spent nuclear fuel, and together with local NGOs spearheaded the organization of a petition signed by the leaders of 90 environment groups that advocated tighter regulation of the program. Interfax, November 25, 2003; and

As a direct result of these efforts, Minatom was forced to justify the details of the first arrangement to import spent nuclear fuel from Bulgaria in fall 2001, as well as to present valid certification for the deal and publicly commit to disbursing 25 percent of the revenues to the Krasnoyarsk budget. This included disclosures of specific transit routes, contract prices, and the role of suspicious offshore intermediaries that the ministry initially had hoped to keep quiet, as they appeared to violate the new legislation and undermine Minatom's early claims of commercial benefit. The public row also prompted Bulgaria to reconsider future spent fuel transactions with Russia, and to solicit bids for construction of a domestic dry storage facility at the Kozloduy nuclear power plant.[87]

Several months later, a similar incident occurred where Russian environmental activists successfully petitioned the Russian Supreme Court to repeal a 1998 presidential decree allowing for the import and burial of 400 tons of nuclear waste from Hungary. Although the deal predated Russia's amended spent fuel legislation, the Court prohibited Minatom from accepting the remaining 377 tons that had not yet been delivered to Russia without formal provisions for returning the radioactive waste byproducts. The ruling obliged Hungary to take back thousands of cubic meters of radioactive waste and encouraged it to pursue alternative plans for expanding its own temporary storage facility. By the end of 2003, the deal was still on hold.

While Putin's institutional revisions left regional leaders impotent to prevent or reverse executive decisions, the governors nonetheless were able to leverage the transparency of their debates to exact greater executive accountability for unpopular nuclear energy policies, thus imposing an indirect constraint on future federal action. By raising specific issues and constraining the terms of debate, regional leaders induced federal agencies to extend additional policy commitments and to incur greater obligations to regional and national constituencies. In doing so, Putin personally assumed greater accountability for potential policy failures. Consequently, regional leaders were able to use their institutionalized standing at the federal level to secure "plausible denial" for problems associated with future implementation. They could criticize the federal government for "pushing through" initiatives, while simultaneously reaping the initial benefits of siding with Putin by officially signing onto his nuclear energy reform agenda. In the event that Putin's reforms become more popular as they unfold, the governors can claim partial credit, as they did not block federal initiatives. By making the policymaking

---

'Chelyabinsk Oblast: The Nuclear Safety Movement Defends Gosatomnadzor', *Regnum-VolgaInform*, January 28, 2003, in *FBIS-SOV* (Document CEP20030313000324).

[87] According to several accounts, Minatom originally charged only $620 per kilo for the 41 metric tons of spent nuclear fuel imported from Bulgaria, in contrast to the $1,000 per kilo that was promised during parliamentary deliberation over the spent nuclear fuel legislation. As a result, several regional and nongovernmental organizations directly petitioned the Federal Prosecutor to ensure reimbursement for the additional $15 million from the deal. In addition, the stipulation to disclose transit routes compelled Minatom to conduct public dialogues with numerous regions in an effort to explore alternative options. *Moskovskiy komsomolets*, October 19, 2001; Interfax, December 4, 2001; and Interfax, July 10, 2003.

process more transparent, regional leaders not only raised public awareness of controversial nuclear energy policies, but situated themselves either to reap the benefits of potential policy successes and/or to render independent, critical assessments of future shortcomings with the federal reforms. As a result, for two-and-a-half years following passage of the legislation, Minatom officials were neither able to land a single new contract nor optimistic about the near-term prospects for breaking into this prospective market.

Finally, the politics surrounding nuclear energy reform clearly demonstrated an emerging horizontal dimension to regionalism in contemporary Russia. As revealed by the two cases discussed in this chapter, the process of regionalism became more complicated than a straightforward devolution of authority from the center to the periphery. With economic and political interests in Russia cutting across federal and bureaucratic frontiers, regionalism was manifest in tactical alliances between the governors and highly institutionalized federal-level interest groups. Faced with reconciling accountability to regional constituencies via formal assaults on federal prerogatives from the center, regional leaders acquired strong incentives to resort to preemptive, informal, and short-term measures—including making side deals with select federal agencies, such as Minatom—to promote their interests. Consequently, Putin's initial federal reforms may have succeeded at arresting the devolution of authority and promoting political integration in Russia—but at a price of limiting Moscow's capacity to impose federal directives or to circumvent the interests of organized regional and economic interest groups within the national policymaking process.

Chapter 4

Minatom's Regional Strategy

Sonia Ben Ouagrham

The Ministry of Atomic Energy (Minatom) stands out as one of the most powerful of Russia's 26 federal ministries. It is responsible for the development, testing, and production of all fissile material, including nuclear weapons, as well as supervising the elimination of nuclear warheads and nuclear munitions and the dismantlement of nuclear submarines. In its post-Soviet incarnation, the ministry's political authority dwarfs that of both the Russian Federal Inspectorate for Nuclear and Radiation Safety (Gosatomnadzor or GAN) and the Nuclear Safety Inspectorate for Nuclear Installations, which are officially charged with regulating civilian and military nuclear sites, respectfully, across the country.[1] Because of its status as a full ministry, Minatom enjoys greater formal standing compared to these subordinated state agencies. In addition, the ministry succeeded at generating significant revenues and at enlarging its political power base via oversight of a host of commercial and administrative responsibilities, such as control of over 150 nuclear production and research facilities, management of dismantlement programs, participation in the sale of blended down highly enriched uranium (HEU) to the United States, and export of uranium products and services.

Notwithstanding its impressive national stature, Minatom's regional posture has been surprisingly inconsistent and in some cases quite lax. On the one hand, officials maneuvered to reinforce the ministry's national political clout and to secure compliance from the regions via formal and de facto control mechanisms. In doing so, they relied on a variety of political, financial, and administrative levers. They also resorted to inducing regional support from targeted public relations campaigns and conditional deals with select territorial administrations. On the other hand, Minatom—under successive leaderships—failed to generate a coherent regional policy. Rather than embracing a concerted strategy for addressing local

---

[1] For instance, Gosatomnadzor lost its power to monitor military facilities in 1995. More recently, in June 2003, Gosatomnadzor influence waned when its outspoken chairman, Yuriy Vishnevskiy, was replaced by Andrey Malyshev, who had been serving as a Deputy Minister of Atomic Energy and is expected to be a less vocal critic of the ministry. While Vishnevskiy officially retired, his retirement as soon as he hit the official retirement age is somewhat unusual. "Kadry," *Kommersant*, June 30, 2003; "Direktor GNIPKI 'Atomenergoproyekt' Andrey Malyshev naznachen zamestitelem ministra RF po atomnoy energii," *RusEnergy - Novosti TEK*, July 19, 2002.

issues and boosting the effectiveness of policies pursued at the regional level, Minatom's actions were informed by the long-term ambition for expanding the nuclear power sector at home and abroad. In this respect, Minatom was prone to treat regions differently, pursuing disparate and *ad hoc* policies oriented primarily towards serving the leadership's grand strategy for expanding construction of nuclear power stations, increasing nuclear energy exports, and consolidating control over the nuclear fuel cycle. While deals were cut with a variety of local leaders, there were few attempts to serve the interests of local populations or to accommodate public concerns for nuclear safety.

Since 1989, when the USSR Ministry of Medium Machine-Building (Minatom's powerful predecessor) was stripped of its mighty construction enterprises and other divisions, the ministry fixated on regaining its lost stature. Yevgeniy Adamov, the minister from March 1998 to March 2001, promoted revenue generating projects, such as spent nuclear fuel imports and was preoccupied with prevailing in political power struggles in Moscow as a means towards securing greater leverage in the federal policymaking process. Consequently, there was less attention devoted to improving the effectiveness of Minatom's regional policies than there was to co-opting influential regional and federal officials, including governors, deputies, ministers, and facility directors. Notwithstanding the real and expected regional effects of these political gambits, Adamov did not seek to address systematically the concerns voiced by local populations and environmental groups, except when they were echoed by regional actors who carried significant political clout. Although the power struggles in Moscow subsided under Adamov's successor, Aleksandr Rumyantsev, Minatom's economic and regional activities nonetheless remained tied to political developments at the federal center. Accordingly, many of Minatom's regional projects remained poorly conceived and inconsistently pursued.

This chapter reviews the basic contours of Minatom's regional strategy and administrative control. The first section outlines the strategy, discussing internal inconsistencies that compromised effective policy implementation below the federal level. The second section describes the political and administrative resources available to the ministry, and analyzes the effectiveness by which they were used to secure regional compliance. The concluding section examines variations in Minatom's regional policies and the range of responses that they engendered.

Minatom's Regional Shadow

Throughout the first decade following the Soviet collapse, Minatom's overarching strategy focused on increasing the ministry's political clout through the promotion of commercial enterprises and the expansion of proprietary holdings. Priority was placed on developing the civilian energy sector, which downgraded attention to the defense sector. This informed the ministry's overall posture, with the exception of those defense transformation-related projects that offered an opportunity to secure additional profits via the dismantlement of nuclear submarines or the sale of downblended HEU to the United States.[2]

The Civilian Nuclear Sector

The development of the civilian sector of the nuclear industry rested on three main pillars. The first was embodied in the *Strategy for Developing Russia's Nuclear Energy in the XXI Century* that was developed between 1998 and 2000. This strategy charted a course for the evolution of the nuclear sector through 2050, with detailed development plans specified for the period leading up to 2010. Specifically, it called for increasing electricity production at nuclear power plants from 21.2 GW in 2001 to 90 GW by 2050, and increasing the nuclear sector's share of total domestic energy production from 14.6 percent in 2001 to roughly 30 percent by 2030.[3] To meet these goals, the strategy required that Minatom construct 30 reactors in 30 years. Initially, it also stipulated the construction of new reactors at nuclear power plants (NPPs) in the European oblasts of Rostov (Volgodonsk NPP), Kursk (Kursk NPP) Tver (Kalinin NPP), and Saratov (Balakovo NPP).[4] This was supplemented by ensuing decisions to build a fourth

---

[2] Through the HEU deal, Minatom has earned some $3 billion. However, in November 2001, the Duma passed a bill providing that the proceeds from the HEU deal be disclosed with a breakdown in accordance with the standard classification of budget items. Since 2001, these funds became a line item of the budget, and therefore subject to oversight by the Russian government accounting office, giving Minatom much less room to maneuver. Profits from submarine dismantlement have been less direct. According to a recent Russian government audit, only 30% of the federal budget funds designated for submarine dismantlement have been spent on priority tasks, while Minatom has misappropriated some $4 million in funds and ineffectively used another $4 million in its efforts to support the creation of a closed nuclear fuel cycle through its submarine dismantlement program. Plans call for Minatom to cede oversight of submarine dismantlement to the Ministry of Economic Development on May 1, 2004. Minatom website, <www.nuclear.ru>, December 3, 2001; Charles Digges and Igor Kudrik, "Audit of Minatom reveals millions in misspent cash and lack of control on sub decommission," Bellona Foundation website, December 5, 2003, <http://www.bellona.no>.

[3] *Yadernaya Rossiya Segodnya*, October 18, 2000; and "Rebro Adamova" (Adamov's edge), *Izvestiya*, November 17, 2000, p. 1. For the revised version of the document, now called the *Strategy for Developing Russia's Nuclear Energy in the First Half of the XXI Century*, presented by Minatom in 2003, see the Rosenergoatom website, <http://www.rosatom.ru>.

[4] An initial investment of $470 million was designated for the construction of the former three reactors, which would provide an additional capacity of 1,000 MW each. The construction of the new NPP in Rostov (designated the Volgodonsk NPP in 2002) started in the mid 1980s, but due to the lack of funding, the project was suspended. In May 2000, Minatom received a license to restart the project, and the construction of Volgodonsk NPP's first reactor was completed at the end of the year 2000. Money to construct a second unit has been designated in Rosenergoatom's 2004 budget. The Kursk NPP already has four reactors and the construction of the fifth is scheduled to be completed in 2006. As far as the Kalinin NPP is concerned, it already houses two reactors, and the construction of the third

unit at the Beloyarsk Nuclear Power Plant (BN-800) in Sverdlovsk Oblast, an additional reactor at Novovoronezh NPP, and a new nuclear power plant in Bashkiriya.

The second pillar of Minatom's commercial strategy targeted the growth in exports of nuclear technology and foreign sales of electricity produced by domestic NPPs. In addition to expanding the nuclear industry at home, Minatom concentrated on building and designing five reactors in China, India, and Iran.[5] Minatom also promoted uranium exports and enrichment services to foreign customers.

The third leg of Minatom's nuclear energy strategy focused on the long-term storage and reprocessing of foreign spent fuel. This dovetailed with efforts to promote the development of a closed nuclear fuel cycle and to facilitate related export projects. However, as discussed in Chapter 3, because Russian legislation in the early 1990s prohibited imports of foreign nuclear waste and did not draw distinctions between radioactive waste and spent nuclear fuel (SNF), Minatom devoted considerable political capital towards amending federal law to advance its commercial projects. As passage of these bills required the votes of regional deputies, Minatom lobbied select regional administrations by proposing to use proceeds from its commercial ventures to finance local ecological programs and submarine dismantlement in the Northwest and Far East regions of the country.[6]

---

reactor is supposed to be completed in 2003. Balakovo NPP already has four reactors. In January 2002, a declaration on the installation of Balakovskaya NPP units five and six, with a capacity of 2 million MW, was signed. Unit five construction is slated to begin in 2004, with completion expected in 2008. Unit six is scheduled to be put into operation in 2010. "Rosenergoatom narashivayet moshnosti" (Rosenergoatom increases capacity), *Yadernyy Kontrol*, May-June 2000, p. 13; "Zelenoye kazachestvo i mirnyy atom" (Green Cossacks and the peaceful atom), *Izvestiya,* June 9, 2000; and Press Conference of Minister Adamov, Minatom website, January 30, 2001, <http://www.minatom.ru/presscenter/document/news>; "Podpisana deklaratsiya o vvode v ekspluatatsiyu vtoroy ocheredi Balakovskoy AES v Saratovskoy oblasti" (Declaration signed on second group of reactors at Balakovo NPP, Saratov Oblast, entering into service), Interfax, January 29, 2002; "Rosenergoatom rasschityvayet, chto pravitelstvo k dekabru utverdit investprogrammu na 2004 god" (Rosenergoatom calculates that the government will approve its 2004 investment program by December), Interfax, November 12, 2003; "Rosenergoatom peredal v pravitelstvo investprogrammu na 2004 god" (Rosenergoatom has given the government its 2004 investment program," Interfax, October 21, 2003; "Rosenergoatom vlozhit v sooruzheniye vtoroy ocheredi Balakovskoy AES 58 mlrd. Rubley" (Rosenergoatom to spend 58 billion rubles on the second set of reactors at Balakovo NPP), Interfax, July 22, 2003.

[5] Press conference of Minister Adamov, Minatom website, January 30, 2001, <http://www.minatom.ru/presscenter/document/news>.

[6] Ibid.

The Defense Sector

Minatom's policies towards the defense sector were governed mainly by the 1998 *Restructuring and Conversion of the Nuclear Weapons Complex for 1998-2000 Program*. The program called for reorganizing nuclear facilities in the closed cities, consolidating weapons production in a limited number of sites, and converting defense-related to civilian activities.[7] This specifically entailed halting manufacture of new weapons at two of the four warhead assembly/disassembly facilities: NPO Start in Zarechnyy and the Avangard facility in Sarov, as well as ceasing production of nuclear weapons components at Seversk by 2004.[8] The program also stipulated that the manufacturing of nuclear weapons was to take place only at the Mayak facility in Ozersk, the Elektrokhimpribor Combine in Lesnoy, and the Instrument-Making Plant in Trekhgornyy. Similarly, Russia's uranium enrichment plants were converted from producing HEU for the weapons complex to producing low-enriched uranium (LEU) for reactor fuel. Additionally, 10 of 13 plutonium production reactors were shut down.[9] The remaining three reactors—one at GKhK in Zheleznogorsk and two at SKhK in Seversk—were slated to continue operations supplying heat and power to local customers and to remain on line until alternative sources of power generation were developed.[10]

Minatom also allocated on average $50 million per year to fund defense conversion projects. In 1999, for example, the ministry used the profits of the HEU agreement with the U.S. to finance 26 conversion programs.[11] This commitment peaked in 2000, when 2.8 billion rubles ($60 million) were allocated from the federal budget to cover the conversion of personnel and restructuring at nuclear weapons facilities.[12] In addition, Minatom created a Department for Conversion in 1999 to coordinate transformation at defense facilities and to target support to those facilities that offered the greatest commercial potential. By the end of the decade, most nuclear facilities had opened new civilian production lines and generally re-directed respective conversion activities towards manufacturing medical technology, physical protection equipment, energy-saving devices, equipment for the agricultural sector, equipment for the energy sector, and electronics for the automotive industry.[13] Minatom also attempted to accelerate the

---

[7] Lev Ryabev, Presentation at the conference on "Helping Russia Downsize its Weapons Complex," Princeton University, March 14-15, 2000.

[8] Oleg Bukharin et al., *Helping Russia Downsize its Nuclear Complex: A Focus on the Closed Nuclear Cities,* June 2000, <http://www.princeton.edu/~cees/arms/conference.shtml>.

[9] Ryabev, "Helping Russia downsize its weapons complex."

[10] The United States agreed to fund the construction and refurbishment of fossil-fueled power plants for this purpose.

[11] Oleg Bukharin, Frank Von Hippel, and Sharon Weiner, *Conversion and Job Creation in Russia's Closed Nuclear Cities,* Princeton University, November 2000.

[12] Ryabev, "Helping Russia downsize its weapons complex." In 2001, Minatom planned to invest about $55 million dollars in conversion and restructuring.

[13] Ibid.

conversion of nuclear facilities by developing the production of equipment and services for the civilian nuclear sector. For example, the All-Russian Scientific Research Institute of Experimental Physics (VNIIEF), located in the closed city of Sarov, initiated development of new safety equipment for nuclear power plants.[14] Finally, Minatom instituted plans to downsize the nuclear sector by 2005, with the objective of reducing aggregate employment within the industry to roughly 35,000–40,000 personnel.

Minatom also enlisted support for defense conversion from large-scale Russian enterprises and regional authorities. In 1999, for example, Minatom signed agreements with the gas monopoly, "Gazprom," for the production of equipment used for oil and gas extraction, as well as with the automobile manufacturer GAZ of Nizhniy Novgorod to produce automobile electronics.[15] Regional leaders also agreed to develop local plans for absorbing the personnel released from the nuclear weapons complex.

Minatom also assumed the lead in overseeing several weapons dismantlement and materials elimination programs, from the HEU-LEU blend-down program to the role of lead coordinator of submarine dismantlement. The former generated significant profits to the closed cities and corresponding facilities, as well as to the ministry itself, as did several other weapons elimination programs. However, not all of these programs posted net gains. Although Minatom acquired marginal profits from submarine dismantlement, it was compelled to invest considerable resources into this effort. That said, the ministry managed to parlay its administrative role into leverage over those regional governments that stood to gain from Minatom's plans to import spent fuel or construct new power plants.

Minatom's Control Mechanisms[16]

In principle, Minatom relied on a set of potent instruments—political, cultural, administrative, and financial—to influence federal policymaking, exercise control over activities and personnel issues at nuclear facilities, and to overcome bureaucratic obstruction within the ministerial hierarchy. In practice, however, these tools generally yielded mixed leverage over regional leaders, local public organizations, or non-governmental organizations (NGOs). They provided only partial influence over the decisionmaking calculus confronting facility directors, who were on the front lines of regional political and administrative networks.

---

[14] "Iskusstvennaya pochka vmesto atomnoy bomby" (Artificial liver instead of atomic bomb), *Vek* 331 (1999).

[15] Ibid. GAZ, in particular, agreed to provide contracts that would result in the hiring of up to 8,000 employees from various nuclear facilities to develop and produce automobile electronics.

[16] This section is based on a paper completed by Igor Khripunov of the University of Georgia for the "Russian Nuclear Regionalism" project. The author is grateful for his contributions to this chapter.

Consequently, Minatom had to complement these efforts with a variety of informal mechanisms aimed specifically at inducing compliance from a range of local actors, including facility representatives, regional leaders, NGOs, and the general public.

Political Control

A primary source of Minatom's influence at the federal and regional levels stemmed from the leadership's direct access to senior decisionmakers, including the Russian president. The minister, for example, regularly participated in meetings held by Russia's Security Council, while other senior ministry officials sat on at least three of the Security Council's interagency departments.The first of these was the Interagency Commission for Defense Industry Security, where Minatom's first deputy minister served as one of the two deputy chairmen. In addition to evaluating defense procurement issues, the commission was tasked with generating priorities for the national technological sector and realizing its export potential. The second Security Council body was the Interagency Commission for Environmental Security, where a deputy minister represented Minatom. The third was the Interagency Commission for Information Security whose membership also included a deputy minister of Minatom. Direct access to these and other senior policymaking bodies allowed the ministry's leadership to participate actively in shaping executive branch policies. The ministry also was authorized to take the initiative in generating recommendations to the government regarding the extension or termination of specific privileges for nuclear facilities. Because of the stature of the ministry, these recommendations tended to provide the basis for government policy.

Financial Control

Another instrument of control derived directly from Minatom's financial discretion. Persistent shortfalls and delays in disbursing payments from the federal budget elevated the ministry's role in providing supplemental funding for regional facilities. The discretion to allocate such funding vested central ministry officials with considerable leverage over nuclear enterprise directors and regional leaders. Hit hard by the travails of economic transition, most nuclear facilities, excluding specific nuclear fuel enterprises, became increasingly dependent on the ministry's coffers.

One source of discretionary funding was generated by the U.S.-Russian HEU/LEU Agreement that averaged roughly $500 million a year. As specified by a 1999 government resolution, Minatom retained the right to compensate those nuclear facilities that produced HEU and down blended it into LEU. It also had authority to allocate residual profits, preferring typically to fund projects related to improving safety at specific facilities, environmental remediation, fundamental and applied research, and defense conversion at specific facilities across Russia. Minatom determined respective priorities for each program and distributed financing to the assigned recipients, even though specific projects were expected to be coordinated with the Finance Ministry and Ministry of Economic Development and Trade. Often, however, coordination was perfunctory, as the generic opacity within the government enabled Minatom to garner nearly exclusive financial discretion over the nuclear industry.

Minatom also was formally obliged to supervise the procurement and coordination of several long-term, nuclear-related programs that were funded directly out of the federal budget. One example was the "Federal Program for Nuclear and Radiation Safety and Security for the Period from 2000 to 2006" that was approved by the government in February 2000. The total budget for this seven-year program was approximately $260 million, with over $200 million earmarked in federal outlays. There were 20 components that encompassed numerous activities: the disposition of radioactive waste and spent nuclear materials; nuclear safety; the safety of nuclear power and research plants; the development of new nuclear power plants with enhanced safety features; training upgrades; physical protection, accounting, and control of nuclear materials; and nuclear safety and security at shipbuilding facilities. The program stipulated that Minatom had to award contracts on a competitive basis and consistent with the federal law "On Tenders for Government Work and Services." Although this law was binding, implementation was considerably opaque. As a result, Minatom wrangled exclusive discretion to grant final awards that were consistent with the leadership's parochial preferences for the ministry.

Administrative Control

Minatom relied on several de facto mechanisms to maintain administrative control over subordinate facilities at the regional level. In practice, Minatom persisted as an unreformed Soviet-style ministry with extensive administrative powers. It maintained direct supervision over many aspects of facility operations, appointing enterprise directors, playing a key role in issuing and supervising state orders and export contracts in the nuclear sector. It also was charged with developing and monitoring technical policies and internal export control mechanisms.

In this context, Minatom's exclusive licensing prerogatives carried considerable political cache. A 1999 government resolution designated Minatom as the licensing and regulatory body for defense facilities dealing with radioactive materials and sources. This applied not only to nuclear weapons technologies and manufacturing, but to naval propulsion systems as well. In the civilian sector, licenses were authorized by Gosatomnadzor but were ultimately issued by Minatom.

Minatom's administrative control over local facilities was reinforced by procedures and a corporate culture forged during the Soviet nuclear sector. Accustomed to the privileged and regimented Soviet structure of administration, personnel across the Russian nuclear complex continued to look to the ministry to preserve traditional values and policies. It was widely accepted across the sector that the survival of the industry depended critically on Minatom's aggressive intervention to coordinate the activities among its territorially and functionally differentiated constituent elements. Accordingly, the ministry enjoyed significant authority to supervise operational and personnel issues at each facility. Senior managers had to be certified by Minatom to maintain their leadership positions and eligibility for promotion.[17] Accordingly, these managers were required to pass up to 15 technical examinations that were jointly administered by GAN and Minatom. Failure to pass the examinations provided grounds for removal from top managerial jobs. Furthermore, directors of both state owned facilities and joint stock companies also had to have their managerial proficiency certified. Minatom was authorized by federal law to provide both training and certification at such facilities, regardless of their proprietary status in order to ensure uniform and consistent standards for operating all Russian nuclear facilities.

Minatom also supervised facility directors through a unique and tightly controlled system for enforcing protection of state secrets administered jointly with other federal agencies. Failure by nuclear scientists or technicians to obtain a clearance or clearance extension could jeopardize their professional careers in the industry.[18] Similarly, the ministry played important roles overseeing the declassification process and approving the clearances for senior managers within the industry to travel abroad. Recommendations regarding the classification of information, technologies, and products were submitted by facility managers, but Minatom wielded final authority to render judgement. Similarly, the process of re-classification was initiated by facility managers but were ultimately approved by the central staff at Minatom. This authority vested Minatom with a powerful political lever, as no product or service could be exported or integrated into a civilian project without obtaining proper declassification.

---

[17] In general, Minatom's leadership certification is valid for three or five years. Only those employees who have occupied managerial positions for over 10 years are eligible recipients. Certificates are issued upon completion of week-long training sessions. Syllabi, developed by Minatom's Institutes of Professional Training and approved by Minatom, are sent to prospective attendees one month in advance. During the training, attendees are requested to submit a research paper on a mutually agreed subject and take an exam focused on relevant nuclear security and safety issues. An appropriate examination commission is set up by Minatom to evaluate attendees' performance. Certificates are denied if an aspiring person is known to have committed disciplinary or other professional infringements, has poor health, or failed the security and safety exam. Leadership certification is becoming an important tool for Minatom to control promotions, manage its top personnel, and improve the quality of its leadership.

[18] Government resolution No. 1050, dated October 28, 1995, lists requirements and procedures for granting clearance and provides samples of registration cards, certificates, and other documents maintained by Minatom. There are three levels of classification: a) information of special importance; b) top-secret information; and c) secret information. Clearance of the first two levels requires approval of the Federal Security Service (FSB) in coordination with Minatom after a thorough background check. Only clearance for the third lower level can be granted by the administrative director of a nuclear facility.

Minatom also was charged with orchestrating inter-agency supervision of Russia's nuclear industry. In the early post-Soviet period, this duty was performed by a subcommittee of the Technical Commission of the State Committee for Standards (Gosstandart) that possessed responsibilities for managing quality control and certification in the nuclear industry. In January 1999, a dedicated certification system for the nuclear industry was introduced that conferred joint responsibility to Minatom, Gosstandart, and GAN. The central working bodies include certification expert centers and testing laboratories. Minatom's major departments were supposed to employ quality assurance experts who would work with the respective entities in the field. Failure to receive proper certification threatened—at least theoretically—to halt operations at a facility. Similarly, Minatom was authorized to establish (together with the Ministry of Industry, Science and Technologies) economic efficiency standards, supervise the management of federally owned property by specific nuclear facilities, and determine the share of revenues to be paid to the federal budget and the amount of dividends to be paid to shareholders. The was applicable to unitary, state-owned facilities as well as to open joint stock companies, if more than 50 percent of the latter's stocks were federally owned.

Minatom also had authority to manage the disposition of federally owned nuclear materials. The ministry served as the custodian of the national stockpile of fissile material inherited from the Soviet nuclear weapons program. Accordingly, Minatom was responsible for overseeing the inventory of the federally owned nuclear materials and reporting its assessments to the State Property Relations Ministry for inclusion in the federal data bank. After 1991, there were attempts to privatize part of the stockpile and to allow nuclear entities to buy uranium at prices significantly below market value. It was not until 1994 that the government approved a set of procedures for direct sales from the stockpiles, with Minatom acting as its primary agent.[19] Since the bulk of the stockpile was mined during the Soviet period, prices were not set on the basis of production costs. Instead, Minatom retained discretion to set prices for each deal that served as another lever to reward or punish a given facility.

In addition to regulating and controlling the performance of facilities and many joint stock companies in the industry, Minatom was authorized to review and finalize draft charters of these facilities. With regard to open joint stock companies where the majority of shares were state owned, Minatom played critical roles brokering cooperation with the Ministry of Property Relations, shaping the agenda of shareholders' meetings, and proposing candidates for boards of directors and auditing commissions. Although the Ministry of Property Relations was appointed as the lead agency, it relied significantly on Minatom's expertise and recommendations.

Minatom's administrative responsibilities extended to the realm of supervising export controls. Before a nuclear facility could enter into negotiations for the

---

[19] After nearly a decade following the Soviet collapse, the total volume of Russia's dwindling stockpiles is estimated at 240,000 tons.

export of products or services, it had to secure approval from Minatom's Export Council, which served as an in-house filter for such operations. Minatom stipulated that any application submitted to the council had to be approved by the ministry's designated internal compliance officer posted at each facility. Designated laboratories would then evaluate export proposals from a technical point of view and issue final reports to Minatom. The Export Council generally considered draft agreements in two phases. First, they were evaluated by experts from the relevant Minatom departments (Department of Information and Facilities Protection, Department of Legal Issues, and International Relations Department). The conclusions by these agencies were then reviewed at a meeting of the council chaired by the Minatom first deputy minister. The authority to approve a contract served as a powerful lever in the hands of the ministry.

Minatom Outreach

In practice, the formidable instruments mentioned above succeeded in preserving Russia's nuclear sector. On occasion, however, the process failed to detect or discipline the shirking of rogue facilities. Such gaps in oversight were due mainly to administrative problems within the ministry. First, Minatom remained divided into several opposing camps that hampered the coordinated application of various control mechanisms. In addition, Minatom's leverage was wielded primarily over facilities, and had only marginal effect on the decisions by regional leaders or local interest and civic groups. Yet, many facilities both openly and discretely appealed to regional/federal leaders, the media, and NGOs either to oppose or circumvent Minatom's projects. In the face of such opposition, Minatom relied on new tools of persuasion. Regarding its projects in the civilian field, Minatom launched a massive public relations campaign aimed at presenting its program in a favorable light, responding to the criticism leveled by regional opponents, and promising financial gains and environmental assistance to various regions. In doing so, the ministry often was able to secure the support of many regional actors that had previously opposed the projects.

Regional Responses

Minatom plans for the civilian nuclear sector were endorsed primarily by those regional actors that derived benefits, financial or otherwise, from the implementation of proposed policies. In return, Minatom acquired additional support from those regions that looked to the ministry for assistance in other areas, such as environmental remediation or submarine dismantlement. Minatom also successfully co-opted regions, such as Rostov Oblast, to secure support for its restructuring program, but was unable to persuade other regions, such as Khabarovsk Kray, which calculated that the environmental risks outweighed the profits derived from the construction of NPPs and the ministry's side payments.

The designers, builders, and contractors designated to implement specific projects also consistently supported the ministry's policies. They stood to become direct beneficiaries of the ministry's efforts to increase its clout and economic weight at the federal level. For instance, the directors of Machinostroitelniy Zavod (in Elektrostal), the producer of nuclear fuel for 70 percent of the Russian fuel rod market (MSZ fuels VVER-440, VVER-1000, AST-500, RBMK-1500, BN-600, EGP-6, and naval propulsion reactors), endorsed Minatom's program for expanding power generation because it offered to increase the number of orders placed at the enterprise.[20] Similarly, the leadership at turbine-producer, Electrosila in St. Petersburg, endorsed the program to boost civilian nuclear technology exports, as the plant received orders for the Bushehr NPP and was slated to service new contracts from China and India.[21]

Similarly, regional administrations that stood to benefit directly from Minatom's programs served as important allies to advance specific projects. For example, former Krasnoyarsk Governor Alexander Lebed supported the spent fuel import and reprocessing project, since the temporary storage facilities for the project were slated to be located in his region and thus stood to boost local revenues and employment opportunities significantly.[22] Another regional champion of the spent fuel import legislation was Kamchatka's State Duma deputy Vice-Admiral Valeriy Dorogin, who made a direct connection between Minatom's SNF profits and expenditures on submarine dismantlement facilities that would take place in his region.[23]

Minatom also relied on traditional supporters to lobby its strategy at the local level. One of the ministry's main allies in the civilian nuclear field was the Union of the Russian Territories and Enterprises of the Nuclear Energy Sector.[24] Created in June 1998, this association became committed to "unifying forces and actions of nuclear energy territories and enterprises to solve social and economic problems in the regions." Accordingly, it organized regular meetings of regional leaders, nuclear facilities' representatives, and Minatom officials to review the status of the nuclear sector and generate policy recommendations.[25] The governors of regions belonging to the union also proved to be invaluable allies in the upper chamber of parliament, as they endorsed several legislative initiatives proposed by Minatom to improve the economic health of the nuclear energy sector. This group of governors

---

[20] *Yadernaya Rossiya Segodnya*, October 18, 2000.

[21] "Peterburg exportiruyet produktsiyu high tech" (St. Petersburg exports high-tech products), *Chas Pik*, No. 15, April 11, 2001.

[22] Nikolai Sokov, Center for Nonproliferation Studies, Monterey Institute of International Studies, unpublished internal report.

[23] Yelena Fedorchenko, "Komu dokhody, a komu – otkhody" (Some profit, some get waste), *Novaya kamchatskaya pravda*, January 25, 2001.

[24] This association includes the following regions: Voronezh, Kursk, Leningrad, Murmansk, Saratov, Sverdlovsk, Smolensk, Tver, Ulyanovsk Oblasts, and Chukotka Autonomous District.

[25] "Yedinye tseli, obshchiye usiliya" (Common goal, common efforts), *Atominform* 41, November 18, 1998, p. 12.

also backed key amendments to a federal law requiring enhanced funding for dangerous radioactive and nuclear production enterprises that was eventually passed by parliament.[26]

In addition, Minatom obtained support from regions that suffered recurrent electricity shortages and that were embroiled in conflict with RAO YeES (*Rossiskoye Aktsionernoye Obshchestvo "Yedinaya Energenicheskaya Systema"*), Russia's electricity distributor. As discussed in Chapter 3, the Chelyabinsk administration began to actively promote construction of the South Urals NPP at the end of 1998. Although construction of the plant was suspended in early 1990 in response to a local referendum, it was resumed shortly thereafter with the support of the regional administration.[27] After the August 1998 Russian financial crisis, which boosted domestic industrial production, manufacturers in Chelyabinsk Oblast faced an electricity shortage.[28] Simultaneously, a conflict between the Troits hydroelectric station—the oblast's main energy producer—and the regional authorities ensued. This was precipitated by the unwillingness of the power station administrators controlled by RAO YeES to provide electricity due to payment arrears. Regional authorities unsuccessfully attempted to seize control over the power station in order to provide local enterprises with a regular electricity supply. Regional administration joined manufacturers in an attempt to reduce dependence on RAO YeES by supporting the construction of a NPP that stood, in turn, to increase its reliance on Minatom.[29]

Notwithstanding the potential economic benefits, Minatom's plans did not receive unconditional support from regional actors. Environmental groups, Minatom's primary opponents in the regions, were especially obstinate in Rostov, Chelyabinsk, Tomsk, and Krasnoyarsk. These groups enjoyed especially strong media and popular support, and aggressively protested against the detrimental environmental impact of Minatom's plans. Other representatives from regional administrations and facilities that stood to lose from Minatom projects also engaged in direct or passive resistance. For instance, in March 2000, Chelyabinsk officials implored President Putin to use his veto should the Duma allow imports of spent nuclear fuel for long-term storage and reprocessing.[30] Similarly, many NNP directors who officially supported the development of Russia's nuclear energy remained suspicious of Minatom's motivations. Anxieties were exacerbated by the

---

[26] "Yedinye tseli, obshchiye usiliya," p. 12.

[27] Natalya Mironova, "Yuzhnouralskaya NPP—A Chronicle of 1992," *Soceco Agency Newsletter*, No. 41-b, November 15, 1992, pp. 1–10; and *Komsomolskaya pravda*, January 26, 1993, p. 3.

[28] As imported goods became too expensive, consumers redirected their demand towards domestic goods that were more affordable.

[29] Similarly, the completion of the Rostov NPP meets a regional demand as this region suffers a strong deficit of energy. Heating shortages in Tomsk-7 are also a major regional problem and the construction of the AST-500 is welcome by the local population. "Zelenoye kazachestvo i mirnyy atom" (Green Cossacks and the peaceful atom), *Izvestiya,* June 9, 2000, <http://win.www.online.ru/rproducts/Izvestiya-Izvestiya-year/09-jun-00/>.

[30] *Yadernyy Kontrol,* No. 3 (May-June 2000), p. 8.

fact that Minatom's strategy for developing the nuclear civilian sector called for a reorganization that would transfer financial control of the NPPs to the ministry. These motivations seemed to be reflected in Minatom's attempt to subsume the Leningrad NPP into Minatom's energy concern, Rosenergoatom, given the NPP's potential for earning significant revenues from exporting energy.[31] In this instance, the regional administration supported the NPP directors in their passive opposition because, under Minatom's planned reorganization, much of the tax revenue from the local NPP was intended to be reallocated to the ministry at the expense of local coffers.[32]

Regional responses to Minatom's plans in the defense sector were even more critical. In general, local facility directors and regional leaders were dissatisfied with the guidance that the ministry provided for defense conversion. There was general frustration with the ministry's preference for short-term, *ad hoc* policies towards defense restructuring, which sent contradictory and ambiguous signals to defense facilities regarding their status within Russia's defense complex. As a result, few facilities actually engaged in substantive restructuring. Conversion programs were often used by default as a means to retain personnel until defense orders resumed. Very few enterprises actually separated civilian and defense activities (personnel, buildings, budgets, and accounting). At Snezhinsk, for example, the Spektr enterprise that was created at the end of the 1990s at the All-Russian Scientific Research Institute of Technical Physics (VNIITF) closed its doors in August 2000 mainly because the director could not afford to permit employees to work full-time on conversion projects. Instead, he implemented a rotation system to ensure that a greater number of employees could obtain additional income. This procedure, however, was initially criticized for its inefficiency and for systematically sapping the incentives to improve productivity.[33]

Regional administrators also complained about the lack of funding for conversion programs. By October 1999, only 25 percent of the annual funds allocated for conversion and restructuring programs were actually received. At other facilities, such as in Snezhinsk, financing of conversion programs was either sharply decreased or not received at all. Typically, funds initially allocated for conversion were reprogrammed for other purposes. In 2000, for example, Minatom planned to provide Snezhinsk with 2.93 million rubles for the development of equipment for the civilian nuclear sector. By the end of the year, however, none of

---

[31] Nuclear power plants in Russia are controlled by Minatom's agency "Rosenergoatom." However, each nuclear power plant manages its own budget independently of Minatom. The Leningrad NPP is the only nuclear power plant in Russia that is not controlled by Rosenergoatom.

[32] "Atomnyy generator idey. Yevgeniy Adamov sozdayet yestestvenuyu monopoliyu" (Atomic generator of ideas. Adamov creates a natural monopoly), *Segodnya*, October 14, 2000, p. 5. The NPP eventually lost the political battle and was integrated into Rosatomenergo.

[33] By the end of 2000, Spektr-Konversiya was completely separated from VNIITF and has produced several successful conversion projects.

these funds were received. Lacking clear guidelines and proper funding, defense facilities often failed to undertake conversion, resisting consolidation to avoid having to compete for survival.

Conclusion

Minatom's activities in the regions have not been informed by an overarching strategy, but instead have been the byproducts of policies pursued in other areas. In particular, Minatom has focused on expanding its civilian sector and maximizing revenue generation in the defense sector. The ministry has resorted to using several direct and indirect mechanisms to affect its policies in the regions, ranging from administrative tools to offering lucrative financial incentives. There has been considerable variation in the way regions interact with Minatom. If a region can assist Minatom with meeting its long-term goals for nuclear power and establishing a closed fuel cycle, it is given every incentive to cooperate with the ministry. Regions that require assistance—whether with weapons dismantlement or environmental remediation—have been more likely to approach Minatom in support of civilian projects than vice versa. Policies, therefore, are not chosen for their effectiveness in redressing specific region issues; they are pursued on the basis for how they advance the ministry's overall long-term strategy. Although Minatom is clearly striving to expand its holdings, the absence of a coherent regional strategy has been counterproductive at times. However, Minatom has faced difficulties in the past and emerged stronger. Thus, only time will tell if it can maintain the support from the regions that it needs to pursue its national goals.

Chapter 5

The Military, the Regions, and Nuclear Weapons

Michael Jasinski

The break-up of the Soviet Union created concern that this unprecedented geopolitical change would result in the emergence of several new and potentially unstable nuclear weapons states. Fortunately, these fears were quickly assuaged by Ukraine's, Belarus's and Kazakhstan's willingness to transfer all nuclear weapons on their territories to the Russian Federation and eliminate all of their strategic nuclear delivery systems and associated infrastructure. But instability in Russia in the early to mid-1990s led to the new threat that Russia itself might not be able to control its nuclear arsenal. This fear was caused by a number of emerging phenomena in the Russia Federation, including: the growth of separatism among Russia's constituent republics (the most visible and violent manifestation of this was the war in Chechnya); an increasingly serious struggle between regional governments and the central government over regional economic autonomy; and the expansion of organized crime.

This chapter first identifies the reasons for the rise of regionalism as a factor in control of Russia's nuclear forces. It then addresses and evaluates the main forms of influence regional actors exert on the Russian military's control of its nuclear forces. These mechanisms range from the possibility of a seizure of nuclear assets by separatists, to the side effects of political and financial conflicts between the center and the regions on the status, financing, and manning of nuclear forces, to the threat posed by penetration of the Russian military by criminal groups.

Decentralization and Emerging Military Regionalism

The rise of regionalism as a factor influencing the Russian military's control of its nuclear forces was to a significant degree caused by the institutional changes experienced by the Russian Federation in the 1990s. In the wake of the break-up of the USSR, the nascent central Russian government proved unable to exercise the same amount of centralized control over regional governments as its Soviet predecessor did. During Boris Yeltsin's presidency, regional governors were able to expand the degree of their autonomy considerably. Concessions by the federal government to the regions included permitting direct elections to the posts of oblast and kray governors and concluding bilateral agreements between the center and regions on taxation. Accompanying the growth of power and influence of regional political leaders was a deterioration in central control of the armed forces, although the change was far less extensive than the political changes occurring simultaneously in the civilian sector. Although the Soviet-era Main Political Directorate was in itself a source of corruption and a debilitating factor in the development of the Soviet military, it nevertheless represented an effective means of centralized political control of the military. Furthermore, the military became increasingly active politically in the late 1980s, a process that also led to its factionalization. Political allegiances of individual senior officers were a significant factor in determining the outcome of the events in Moscow of August 1991 and September 1993.

It should also be noted that a major contributor to the rise of regional influences as a factor affecting Russia's nuclear forces were the severe financial problems facing the Russian military, including even its most elite and sensitive components. These perennial financial shortages have made the military much more susceptible to regional influences. In fact, these problems have not only forced the military to acquiesce in regional and local governments' encroachment on its prerogatives, but in many instances have resulted in the military actually approaching regional governments with requests for assistance. Although the military's financial situation began to improve in the late 1990s and problems with timely payment of military salaries have now been greatly reduced, the social situation in the armed forces is still far from optimal.

Some 46 percent of Russian servicemen and their families still live below the official poverty line, as compared to 33 percent of the civilian population.[1] Although military salaries were increased in 2002, this increase was accompanied by a cancellation of a number of benefits and privileges, thus reducing the actual impact of the pay raise.[2] For instance, whereas a heavy bomber division commander's monthly salary increased from 8,000 rubles to 12,000 rubles, this increase was accompanied by a cut in qualification pay.[3] Moreover, this overall 10 percent pay raise did not even keep pace with inflation, which, according to official data was 15 percent in 2002. As a result, while the average Russian citizen's wages increased by 35 percent during 2002, the military's real wages decreased by 3 percent. Moreover, the average monthly salary for hired labor exceeded pay offered to contract soldiers.[4] As a result of its failure to offer salaries competitive with the civilian market, the Russian military's attempts to attract contract soldiers

[1] Vladimir Temnyy, "Ne v toy zhizni," Grani.ru website, February 9, 2001, <http://www.grani.ru/>.

[2] Vladimir Temnyy, "Komandovat paradom budet Nikolayev," Grani.ru website, June 15, 2001, <http://www.grani.ru/>.

[3] Olga Bozhyeva, "Strategicheskaya spiral," *Krasnaya zvezda* online edition, August 7, 2002, <http://www.redstar.ru/2002/08/07_08/1_01.html>.

[4] Vladimir Temnyy, "Komandovat paradom budet Nikolayev," Grani.ru website, June 15, 2001, <http://www.grani.ru/>.

to perform certain more demanding jobs have resulted in difficulties in attracting qualified personnel. For instance, the salary of contract security personnel guarding the Vladimir Strategic Rocket Forces (SRF) army headquarters ranges between 1,300 and 1,750 rubles ($40-$55). This level of pay has proved too low to attract men to the job and, as a result, the security force has been forced to accept female recruits.[5]

Apart from struggling to provide adequate pay, the Russian government has not been able to address other social issues, even in the strategic nuclear forces. SRF units have been trying to cope with the problem of providing housing for retired personnel. As of the end of 2000, 16,000 SRF personnel had no apartments, and 21,000 retired SRF troops were still living in the closed garrisons awaiting housing that was due to them by law. In many cases, discharged officers have to wait for up to 10 years until they receive an apartment.[6] By mid-2002, the situation had improved somewhat, with the number of contract SRF personnel without housing dropping to 14,500, or 28 percent of all SRF officers and contract enlisted soldiers, and the number of discharged SRF personnel still occupying military housing declining to 14,300. In spite of that improvement, the situation was far from being resolved, since between 40 and 80 percent of housing (depending on the specific garrison) was occupied by discharged personnel for whom the SRF was unable to find apartments outside of the military garrisons, and the amount of money spent on housing was less than in preceding years. While the SRF received $2.7 million to procure new housing in 2000, in 2002 that amount was reduced to just $700,000. An attempt to reduce the housing problem by issuing housing certificates failed due to the difference between the value of the certificates and actual market prices for apartments.[7] By June 2003, the number of SRF officers without housing had decreased to 13,500, although that decrease may have been due to the overall decrease in SRF personnel strength, rather than a net improvement in the number of available apartments. As the SRF expected to receive only 1,800 new

---

[5] Yevgeniy Nizov, "Yadernyy shchit strany derzhat vo Vladimire," *Argumenty i fakty*, December 4, 2002; in "Vladimirskoye obedineniye raketnykh voysk strategicheskogo naznacheniya - odno iz elitnykh podrazdeleniy rossiyskoy armii," Novosti-online, December 4, 2002; in Integrum Techno, <http://www.integrum.ru/>.

[6] Andrey Matyakh, "Topical Subject," *Krasnaya zvezda*, December 23, 2000, p. 5; in "Russian Strategic Missile Troops Military Prosecutor on Servicemen's Rights," FBIS Document CEP20010105000116.

[7] Aleksandr Vovk and Aleksandr Dolinin, "Lyudi i raketi: sotsialnyy aspekt," *Krasnaya zvezda* online edition, September 5, 2002, <http://www.redstar.ru/2002/09/10_09/2_03.html>; Vadim Koval, "Gde zhit leytenantam?" *Krasnaya zvezda* online edition, September 10, 2002, <http://www.redstar.ru/2002/09/05_09/1_01.html>; Salavat Suleimanov, "Moskva utochnyayet raketnyye prioritety," *Nezavisimoye voennoye obozreniye* online edition, August 23, 2002, <http://nvo.ng.ru/printed/forces/2002-08-23/1_priorities.html>.

apartments in 2003, the housing problem is likely to persist into the foreseeable future.[8]

The situation is identical in other branches of the military. As of November 2002, there were 500 officers without housing in just one strategic bomber division, based at Ukrainka. This situation was due to the delays in finding housing for officers who had been discharged from the Air Force but who still occupied military housing facilities.[9] The Space Forces, which control Russia's strategic early warning assets and the nuclear-capable anti-ballistic missile system near Moscow, were also short 7,000 apartments as of December 2002. Here, too, the rate of acquisition of new apartments is inadequate to the needs. In 2002, the Space Forces acquired 174 only apartments, and in 2003 hoped to acquire only an additional 258.[10]

Even the Ministry of Defense's (MOD) 12th Main Directorate, which is in charge of maintaining and safeguarding nuclear munitions assigned to the MOD, has not been able to avoid the social problems that have been affecting other military organizations. Pay is low, nuclear storage facilities are located in remote and unattractive areas of Russia, and facilities' staffs are allegedly undermanned. Poor social conditions were responsible for the minor incident staged by the wives of 12th Main Directorate officers assigned to a facility near Tula. Frustrated by the government's failure to pay their husbands' wages, the officers' wives briefly blockaded the main entrance to the facility. Following this incident, the facility was closed and personnel serving there were reassigned to Moscow, where makeshift housing was made available to them. Since Russia's nuclear warhead security systems are dependent on the human factor to a greater extent than U.S. systems, anything affecting the reliability of the guard force will have a correspondingly greater impact on the security of nuclear warheads.[11]

Finally, for most of the 1990s, the Russian military was unable to pay for the water, gas, and electricity services provided, in most cases, by regional utility companies. This failure has resulted in some military units accumulating massive debts to utility companies, which on occasion have cut off vital services.

These problems not only contribute to a greater susceptibility and willingness of military formations to accept assistance from local and regional circles. They have also created a severe personnel retention crisis within the military, which especially affects junior- and middle-grade officers. Since the military no longer offers a guarantee of material benefits (including housing), officers who are not within a few years of retirement increasingly opt to seek employment outside of

---

8 Vadim Koval, "S uchetom mestnogo faktora," *Krasnaya zvezda*, No. 121, June 21, 2003; in Integrum Techno, <http://www.integrum.com>.

9 Aleksandr Bogatyrev, "Boyegotovnost-glavnyy kriteriy," *Krasnaya zvezda*, No. 207, November 11, 2002; in Integrum Techno, <http://www.integrum.com>.

10 Agentstvo voyennykh novostey, December 9, 2002; in "Russian Space Troops commander says 7,000 troops need housing," FBIS Document CEP20021209000099.

11 Kirill Belyaninov, "Nuclear Decay," *Novyye izvestiya*, February 19, 2000, pp. 1, 4; in "Questions on Russia's Nuclear Security," FBIS Document CEP20000223000309.

the military, creating a vacuum of skilled professionals at junior- and middle-grade levels of command. This results in the premature promotion of less experienced personnel. The prolonged economic privation experienced by a large segment of the Russian officer corps has also caused a deterioration of morale and discipline. Although these phenomena rarely manifest themselves in a form as extreme as the March 2002 suicide of Lieutenant Igor Boytsov, the officer in charge of a nuclear warhead security detail at the Novosibirsk SRF division,[12] they nevertheless represent a major problem for the Russian military.

Assessing the Risks of Terrorism and Nuclear Separatism

Of the wide range of possible regional influences on nuclear weapons, the threat of seizure by a separatist organization appears to be the most remote. In spite of the severity of the conflict in Chechnya, at no time were Russia's nuclear weapons directly threatened by the Chechen conflict, although in 1996 a group of Chechens was reportedly arrested near Tula for spying on a 12th Main Directorate facility.[13] The Chechens may have also been linked to two attempts by terrorist groups to penetrate nuclear storage facilities (the so-called "S" sites belonging to 12th Main Directorate) in 2001. These incidents were later admitted by the chief of the 12th Main Directorate, General Igor Valynkin, although no additional details were provided. Both attempts proved unsuccessful.[14]

The Chechen conflict, however, has compelled the Russian military to take additional precautions, including the removal of nuclear weapons from the Mozdok air base, located only 10 km from the border with Chechnya. The weapons based there were transferred to Engels air base in Saratov Oblast without any incident. By the end of 1998, all strategic bombers based at Mozdok were transferred to Engels as well. This appears to be the only case of the Russian military moving nuclear weapons in response to a potential terrorist threat.[15] Furthermore, as a result of the Chechen conflict, the importance attached by Russian military doctrine to such missions as combating separatism and terrorism has increased.[16]

Apart from Chechnya, there have also been concerns that weapons stored in regions with a largely non-Russian population or headed by a maverick governor might somehow fall under the control of a local government that might use them to blackmail the central government. Like the threat that the Chechens (or other

---

[12] *Agentstvo Voyennykh Novostey*, March 15, 2002; in "Strategic Missile Forces Serviceman Commits Suicide in Southern Siberia," FBIS Document CEP20020315000040.

[13] Kirill Belyaninov, "Nuclear Decay," *Novyye Izvestiya*, February 19, 2000, pp. 1, 4; in "Questions on Russia's Nuclear Security," FBIS Document CEP20000223000309.

[14] "Minoborony RF: presecheny dve popytki proniknoveniya na sklady yadernykh boyepripasov," Strana.ru website, October 25, 2001, <http://www.strana.ru/>.

[15] William Arkin, Robert S. Norris, Joshua Handler, "Taking Stock," NRDC Nuclear Program, March 1998.

[16] "Russian Defense Minister Warns About Terrorism," Interfax, November 23, 2000.

armed separatist groups) might seize nuclear weapons, the possibility of local political elites attempting to use nuclear weapons as political bargaining chips is also remote. Aside from the question of their ability to storm nuclear weapon storage facilities, the political aspirations of the Chechens and other potential separatist movements would seem to preclude the resort to weapons of mass destruction (WMD). Doing so would threaten to undermine their political legitimacy, especially in the eyes of the world community without whose recognition such movements have few chances of succeeding. Historically, WMD have tended to be used by organizations with no well-defined political goals, for example, Aum Shinrikyo, which was responsible for the nerve gas attack in Japan. However, it should be noted that the Nord-Ost hostage crisis in Moscow in October 2002 has raised the possibility that continuing radicalization of the Chechen resistance may lead them to seek WMD capabilities.

The much-celebrated case of Aleksandr Lebed's 1998 "threat" to seize control of ICBM units in Krasnoyarsk Kray illustrates just how unlikely such a scenario is. In spite of the superficially threatening nature of Lebed's statement, in actuality, it did not reflect a desire on his part to seize a part of Russia's nuclear arsenal. Lebed had made several past attempts to draw attention to the plight of Russia's military personnel and did not appear to have any separatist plans (he had sharply criticized another gubernatorial candidate for making threats to take the oblast out of the Russian Federation). Lebed's hints of the possible loss of control over nuclear warheads most likely represented another attempt to draw attention to the broader problem of the military.[17] Lebed emphasized that the main threat to the security of nuclear forces in Krasnoyarsk Kray was posed by the potential actions of unpaid SRF officers.[18] There are no indications that Lebed ever took any actions to assume control over the nuclear assets on his kray's territory, or tried to use them as a political "bargaining chip" of any kind.[19] Following his death in a helicopter crash in April 2002, no Krasnoyarsk politician has revisited this issue.

The possibility of a seizure of control over nuclear weapons does not appear to be greater when the regional governor is a former military officer (like Lebed). To be sure, recent years have seen large numbers of officers enter political life. However, such officers tend to have extensive ties to the local political and economic elites. Although former senior officers are successful in attracting the votes of military personnel, military support is not as key a factor in their elections as the support of the aforementioned regional elites. The support of regional elites was a key factor in the election of the former Baltic Fleet commander Admiral Vladimir Yegorov to the post of Kaliningrad Oblast governor. Although Yegorov

---

[17] "Lebed warns Kremlin over troops' pay," NUPI Centre for Russian Studies website, July 27, 1998, <http://www.nupi.no/russland/russland.htm>.

[18] "Lebed Offers Kirienko Nuclear Headache," RFE/RL Newsline, RFE/RL website, July 27 1998, <http://www.rferl.org/>.

[19] *Nezavisimaya gazeta*, December 10, 2000; in "Russian paper says Krasnoyarsk governor Lebed seeking political comeback," BBC Monitoring; in Johnson's Russia List #4687, December 14, 2000.

also enjoyed the support of Vladimir Putin and received the vast majority of votes cast by the sizable military contingent in the oblast, it was the support of regional political and economic elites that was the deciding factor. Yegorov's support included Lukoil, the largest taxpayer in the oblast, the mayor of Kaliningrad, and scores of other leading businessmen and politicians of the region. By contrast, his opponent's support came mainly from rural areas.[20]

Some fears regarding regional separatism have been caused by Vladimir Putin's creation of federal districts. These new entities were created to solidify the center's authority over the regions and promote the alignment of local laws with federal laws. Until the merger of the Urals and Volga military districts (MDs), federal district borders coincided with those of the MDs. To govern these seven districts, Putin appointed presidential representatives, several of whom were active-duty military general officers before their appointments, including several with combat experience in Chechnya. Some observers have voiced concern that, far from solidifying federal authority over the regions, Putin was in fact creating new regional structures. By being more viable economically than the smaller and geographically isolated subjects of the federation, these entities raised fears that they might eventually challenge the center and even secede from the federation.

However, this new institution has not had a decisive influence on regional power. In part, this was caused by their vaguely defined responsibilities. General Viktor Kazantsev, the presidential representative to the Southern region, saw his job as coordinating efforts of other government structures, rather than exercising executive powers. He was not even granted control over money flows between the center and the regions, a key measure of authority.[21] When in 2000 military units stationed in the Ivanovo Oblast were disconnected from the power grid for unpaid energy debts and local military units took matters into their own hands, the involvement of Georgiy Poltavchenko, the presidential representative to the Central Federal District that includes Ivanovo Oblast, was limited to praising the unit commander for taking this action to reverse what he called an "arbitrary act" by the utility company and to referring the matter to the procurator's office and to the president. This incident further illustrates the relatively limited authority possessed by the presidential representatives.[22]

It appears that the threat of separatism by Russian Federation subjects has considerably subsided since the mid-1990s. In any event, there seems to be an emerging awareness that the possession of nuclear weapons is more of a liability

---

[20] Gary Peach, "Admiral Voted Kaliningrad Governor," *Moscow Times* online edition, November 21, 2000, <http://www.themoscowtimes.com/>.

[21] Nikolay Poroskov, "Russian President's Southern Federal District Plenipotentiary Representative Viktor Kazantsev: 'I Love Going for a Walk With a Gun and Thinking'," *Vek*, January 11, 2001, p. 12; in "Putin's Southern District Viceroy Kazantsev Interviewed," FBIS Document CEP20010109000134.

[22] Vitaliy Denisov, "Russia Will Have A Better Tomorrow," *Krasnaya zvezda*, February 8, 2001, pp. 1, 2; in "Denisov: interview with Georgiy Poltavchenko, Presidential Representative to Central Federal District: national security issues at regional level, relations with regional governors, military reform," FBIS Document CEP20010207000438.

than an asset. Even in the event of the successful political separation of a Russian region with nuclear weapons on its territory, the scenario would likely follow the path already charted by separatist former Soviet republics that gave up their nuclear weapons in return for international support and recognition. While they used nuclear weapons to gain political capital, they did so by giving them up, rather than retaining them. Nor do threats to seize control of nuclear weapons figure in the arsenal of tools regional governments use in their struggle for power with the federal government.

Regional Economics and the Military

Instead of threats of seizure, regional influences on nuclear forces are more subtle and more closely related to the ongoing struggle between the regions and the center for control of tax revenues and over financial burden-sharing. These influences do not represent an attempt to gain control over the military, but instead are motivated by a wide range of political and economic considerations. This general struggle has affected the status of regional military installations, which, on occasion, have become bargaining chips in negotiations between the center and the regions. At the most basic level, regional governments are interested in retaining military bases on their territories. In addition to the prestige of hosting them, military bases are important economic assets in their own right. They provide jobs for local civilian personnel and purchase a wide range of services from the local economy. It should also be noted that military units themselves frequently turn to regional authorities for assistance. In other words, regional influence on the military does not always take the shape of a regional "push," but frequently of a military "pull" as well.

Regional political and industrial leaders appear to be more concerned about exerting their influence over the law enforcement and investigative agencies of the Russian government, which have jurisdiction over key matters of property ownership, taxes, and other economic levers that are at the heart of the regional elites' concerns. Regional governors and the center also continue to battle over the control of revenue streams within the Russian Federation. As a result of the fluidity in center-regional relations and the inability of the federal government to provide adequate funds for its armed forces, regional authorities have acquired some influence on military planning and decisionmaking, even with regard to nuclear forces. Regional authorities have also rendered considerable economic assistance to military units, including components of Russia's nuclear forces stationed on the territory of their regions.

Since regional elites often see military bases as important economic assets, such bases have on occasion become the objects of political struggles. Examples of military bases being drawn into such struggles include the Engels air base in Saratov Oblast, one of the Russian Air Force's two bases for strategic bombers. In September 2001, the Saratov Oblast Security Council received permission from the Russian Air Force to give the Engels air base the dual status of an international airport and a military base. This idea reportedly belonged to Saratov Oblast

Governor Dmitry Ayatskov, who lobbied the Russian government to implement this measure.[23] This issue was already on the agenda in March 2000, when Ayatskov met with LRA commander General Mikhail Oparin and Deputy Prime Minister Ilya Klebanov to discuss the possibility of using Engels Air Force base for civilian uses.[24] Work on converting the base to dual use began in late 2002, and is to be completed by 2005. The project, with an estimated cost of $300 million, will involve enlarging the territory of the airbase by nearly 150 acres in order to create the necessary civilian infrastructure. However, military and civilian aircraft will share the same runways, navigational and communications equipment, and refueling facilities. In spite of the initial resistance to the project on the part of the Air Force, it now supports the conversion on the grounds that joint use of the air base will reduce the financial burden on the military and also provide funds for resolving housing and social issues, such as finding employment for discharged officers. It is not clear, however, whether the Air Force's hopes will be borne out in practice, since Saratov Airlines, the company that stands to benefit from this project, has stated that it does not bear any responsibility for financing military operations at the air base.[25] Although both the Air Force and the oblast government promised to take measures to preserve the security of the base, the presence of a commercial airport on the territory of a heavy bomber base will inevitably affect its operations. The decision is also an indicator of the leverage some regional governors have over the fate of military bases on their territories, and of the willingness of the federal government to use even base housing components of its strategic nuclear arsenal as bargaining chips in political negotiations with regional leaders.

Another dimension of this issue is the fact that regional political leaders often enjoy good relations with local senior military commanders that are based on shared economic interests. An example of such a relationship is the close cooperation between Sverdlovsk Oblast Governor Eduard Rossel and General Yuriy Grekov, the erstwhile commander of the Urals MD. When Grekov was replaced as the Urals MD commander, he immediately assumed a post in Rossel's administration as the advisor on issues affecting the military-industrial complex.[26]

---

[23] "O voyennom aerodrome v gorode Engelse," Interfax-Agenstvo Voyennykh Novostey website, September 21, 2001, <http://www.militarynews.ru/>.
[24] Itar-Tass Weekly News, March 20, 2000.
[25] Nadezhda Andreyeva, "Pikiruyushchiy aeroport," *Vremya-MN*, October 4, 2002; in Integrum Techno, <http://www.integrum.com>.
[26] Yelena Smirnova, "Why Is Petr Latyshev Fighting Eduard Rossel?," *Nezavisimaya gazeta*, November 3, 2000, pp. 1, 4; in "'Confrontation' Between Sverdlovsk Governor, Putin's Representative Viewed," FBIS Document CEP20001107000042.

Military Dependence on Regional Infrastructure

The struggle between the regions and the center over revenues and financial responsibilities has made Russian military units, including nuclear ones, hostages to that struggle and vulnerable to periodic losses of utility services and benefits. The unequal distribution of tax revenues and services between federal and regional governments has led regional authorities to complain that they do not obtain a sufficiently high proportion of tax revenue to cope with their responsibilities for providing education, health care, and public utilities on their territories. As a result, regional authorities have occasionally attempted to reduce their expenditures by curtailing benefits for military personnel and cutting off the supply of electricity to debtors, including the military.[27] The Russian military's failure to pay electric bills has threatened a wide range of facilities, including nuclear-capable ones, with power shutoffs. In some cases these shutoffs might have had dire consequences. Even when power shutoffs do not result in nuclear incidents, they do nevertheless have a negative effect on the morale of troops, combat readiness, and the safety of nuclear units. To avoid such power shutoffs, many military installations that are unable to pay have resorted to barter or to provision of "in-kind" services, often involving the use of military conscripts as cheap labor. One SRF base had its power shut off after it failed to pay an energy debt that amounted to $683,000. When in 2000 an SRF missile unit stationed in Ivanovo Oblast was disconnected from the power grid for unpaid energy debts, the unit commander took matters into his own hands and dispatched troops to the utility company in order to have the electricity restored. Fortunately, the only areas affected by the shutoff were the missile division's garrison areas, not operational missile units.[28] The SRF division stationed near Drovyanaya (Chita Oblast) has suffered from a lengthy period of daily four-hour shutoffs of electricity. As in the case of other SRF divisions that have suffered from this predicament, only garrison areas of the unit were affected.[29] The conflict between power utilities in Primorskiy Kray and military units (as well as other debtors) reached a new level in February 2001, when a local military commander ordered a call-up of the staff of Dalenergo, the local utility, for refresher military training. The commandant relented when he received an urgent request from the utility company explaining that the call-up had deprived the energy company of too many members of its staff.[30]

---

[27] Sergey Pavosudov, "Life or Money?" *Nezavisimaya gazeta*, January 18, 2001, pp. 1, 3; in "'No Hope' Municipal Budgets Can 'Pay in Full for Electricity'," FBIS Document CEP20010118000247.

[28] Vitaliy Denisov, "Russia Will Have A Better Tomorrow," *Krasnaya zvezda*, February 8, 2001, pp. 1, 2; in "Denisov: interview with Georgiy Poltavchenko, Presidential Representative to Central Federal District: national security issues at regional level, relations with regional governors, military reform," FBIS Document CEP20010207000438.

[29] Interfax, February 5, 2002; in "Russia: Power Cut Continues at Siberian Missile Forces Base," FBIS Document CEP20020205000048.

[30] ORT, February 28, 2001; in "Energy debt recovery, evasion reaches bizarre proportions in Russian Far East," FBIS Document CEP20010228000300.

Compared to the electricity crises that have affected SRF bases, shutoffs of naval bases that are home to nuclear submarines could have even more serious consequences. Nuclear submarines require an external energy supply when they are in port with their reactors powered down. The external power supply is needed not only for normal operations but also to operate reactor safety systems. In at least one case, an abrupt power cutoff nearly led to a nuclear accident aboard a submarine.

The struggle between the center and the regions over revenue and burden-sharing has also extended to the area of military personnel benefits. Russian military personnel are entitled to a wide range of services, many of which are funded by regional authorities. As a result, military servicemen have become a burden on local budgets and many regional authorities have deprived servicemen of their benefits. In 2000, the Chita City Duma forbade military personnel from using public transport free of charge, thus requiring intervention by the local military prosecutor. A similar case took place in Astrakhan Oblast, which required payments from discharged soldiers requesting a medical examination. The SRF Prosecutor's Office was compelled to intervene to restore the benefits to some 1,500 troops and retirees when the local telephone company refused to give them military discounts.[31] Some 300,000 military personnel had their rights infringed upon by local governments in such a fashion in 2000 alone.[32]

It appears that the conflict over free benefits is in the process of being resolved, albeit in favor of regional governments. In 2001, military personnel began paying income tax. Part of the justification for having the military pay income tax was to compensate local authorities for the use of services on their territories.[33] Secondly, the military's privileges are being reduced. On November 28, 2000, Chief-of-the-General-Staff Anatoliy Kvashnin announced that social privileges would be replaced by monetary compensation. Some 70 percent of Russia's population enjoys special privileges of one kind or another, and reducing their number is seen as necessary to relieve regional authorities of various financial burdens. Military and law enforcement personnel will be the first to undergo the reform.[34] It remains to be seen whether the government will be able to compensate its military personnel.

---

[31] Andrey Matyakh, "Topical Subject," *Krasnaya zvezda*, December 23, 2000, p. 5; in "Russian Strategic Missile Troops Military Prosecutor on Servicemen's Rights," FBIS Document CEP20010105000116.

[32] Mikhail Timofeyev, "The Army's Defenders of Rights," *Nezavisimoye voyennoye obozreniye*, January 26, 2001, pp. 1, 3; in "Chief Military Procurator Kislitsyn interviewed on law and order, crime issues in the military in 2000," FBIS Document CEP20010129000380.

[33] Aleksandr Lyutov, "Armiya nachinayet platit nalogi," Grani.ru website, January 31, 2001, <http://www.grani.ru/>.

[34] Natalya Ilina, "Awkward Privileges," *Segodnya*, November 29, 2000; in "Social Reform Depends on Military Privilege Abolition," FBIS Document CEP20001129000189.

Regional Economic Assistance to the Military

As a result of the federal government's inability either to finance the existing armed forces adequately or carry out deep force cuts to bring Russia's force structure more in line with its financial capacity, the proportion of Russian military spending being borne directly by the regional governments steadily has grown. The federal government appears to have acquiesced in this trend, since local funding of the military poses a lesser threat than the absence of such funding. Moreover, in many cases military units themselves have turned to regional authorities or even businessmen with requests for assistance. In some cases, such regional contributions are done through an arrangement with the federal government. For example, Russian regional governments share the cost of the semi-annual call-ups of new draftees with the federal government. In the fall 2000 call-up, the regions spent twice as much on the call-up ($1.9 million) as the federal government did ($900,000). Funds provided by the regional government were used to pay officials and health care personnel involved in the call-up and covered the cost of transportation of new recruits to their assembly points.[35]

In other cases, such aid is rendered on an *ad hoc* basis without consulting the federal government. Some city and regional governments have begun to sponsor military units, frequently ones not located on their territories. The network of such sponsorship arrangements is quite wide and includes a large number of cities and military units, both conventional and nuclear. For example, the Orenburg-based SRF missile army HQ maintains close ties with local authorities and the mass media, and the oblast has assisted the army in organizing military-patriotic education of pre-draft age youth. There is also some economic cooperation and assistance between the two entities. There has been some controversy concerning the transfer of the army's social and cultural facilities to the city, particularly the army's need for assurances that servicemen will be able to use the facilities after their transfer.[36]

Some other regions have taken under their "patronage" Tu-95MS heavy bombers, such as those stationed the Ukrainka air base in Amurskaya Oblast. In return for his support, Irkutsk Mayor Vladimir Yakubovskiy became an honorary member of the flight crew of the bomber bearing his city's name.[37] It is not unheard of for one city to sponsor more than one military unit. The city of Tambov has patronage agreements with a nuclear submarine, a border guard unit, and a heavy bomber unit, which named one of its Tu-95MS bombers after the city. The importance of such arrangements to the military was underscored by General

---

[35] Agentstvo Voyennykh Novostey, February 19, 2001; in "Russia Summarizes Results of Autumn 2000 Draft," FBIS Document CEP20010219000202.

[36] Aleksandr Dolinin, "The Best Remain in Formation," *Krasnaya zvezda*, December 7, 1999, p. 2; in "Interview with Head of Orenburg RVSN," FBIS Document CEP19991206000045.

[37] Viktor Otinov, "V nebesakh, na zemle i na more," *Vostochno-sibirskaya pravda*, November 2, 2000; in Integrum Techno, <http://www.integrum.ru/>.

Mikhail Oparin's visit to Tambov for the ceremony.[38] Sometimes a local government's aid to a military unit may be in the form of a one-time benefit. In response to pleas for assistance by the Urals Air Force and Air Defense Grouping commander, Lieutenant General Yevgeniy Yuriyev, in October 2001, the Yamal-Nenets Autonomous District procured a large quantity of petroleum products for the unit. Yuriyev felt compelled to request assistance from local authorities because his unit had received only 30 percent of its minimum fuel allowance.[39] In other cases the relationship between military units and local and regional governments is more symbiotic in nature. In an article published in the Kaluga newspaper, *Vest*, shortly before his reassignment to Moscow, Kozelsk SRF Missile Division Commander Major General Sergey Karakayev praised local authorities for their understanding and cooperation in helping the division address its social needs. Karakayev noted that the city also benefited from having the 6,000-strong division stationed in Kozelsk, which has fallen on hard economic times. With most of the businesses in and around the city inactive, the division has become the most important employer in the area.[40] Similar mutually beneficial cooperation is taking place between Irkutsk Oblast authorities and SRF units stationed in the oblast.[41]

A particularly illustrative example of what such sponsorship arrangements can entail took place in October 2000, when representatives of several Russian cities, members of the "Goroda Rossii" association, met in Vilyuchinsk, Kamchatka, with the commanders of submarines they sponsor. Cities represented in the meeting included Vilyuchinsk, Zelenograd, Irkutsk, Krasnoyarsk, Omsk, Petropavlovsk-Kamchatskiy, Tomsk, and Chelyabinsk. The goal of the visit was to work out ways for rendering services under the sponsorship agreements, and to coordinate the efforts by various cities, with the goal of enhancing the combat readiness of sponsored units, improving personnel welfare, and boosting the prestige of military service. One of the issues on which military and city representatives disagreed was the degree of involvement of cities in the lives of soldiers. Some cities had extensive assistance programs in mind, including the provision of telephones, newspapers, construction or repair of housing, and even addressing the living conditions on the ships, as well as funding training. Furthermore, the cities' representatives wanted the sponsorship to be regulated by a legally binding document, which would place conditions not only on the cities, but on the military units as well. The city delegations signed an agreement on October 12, 2000, concerning development of sponsorship relations between the cities and the Pacific Fleet. Under the agreement the cities would conduct military-patriotic work among

38 "Teper Tambov ne tolko plavayet, no i letayet," *Komsomolskaya pravd–Voronezh*, July 27, 2000.

39 "Pomoshch ot gubernatora," *Krasnaya zvezda* online edition, October 23, 2001, <http://www.redstar.ru/>.

40 Sergey Karakayev,"Raketnaya krepost," *Vest*, July 11, 2001; in Integrum Techno, http://www.integrum.ru/>.

41 "Oblast pomozhet raketchikam," Teleinform-Irkutsk, June 8, 2001; in Integrum Techno, http://www.integrum.ru/>.

pre-induction youth (therefore competing with ROSTO, the federal institution tasked with conducting such activities), and recommend for fleet service the best candidates identified during pre-induction training. This would provide the local city authorities with considerable authority to determine the Navy's personnel policies and reduce the military's authority by bypassing military selection processes and removing its flexibility in personnel decisions. The agreement also requested that the Pacific Fleet commander require the fleet to send draftees from sponsoring cities to the ships they sponsor. In addition to intruding into personnel issues, this sponsorship affects issues of military procurement. The city of Omsk, for example, offered to equip its sponsored warship with locally manufactured rescue apparatus and other equipment.[42]

Similar arrangements exist in other branches of service. One of the regiments belonging to an SRF division headquartered in Tver has been accepting conscripts only from Tver oblast since 1999. The government of the oblast has also provided supplies of food, medicine, and reading materials to the unit. Remarkably, Russian Defense Minister Sergey Ivanov has praised this arrangement as an example to be emulated.[43] This is not the only such example. The Yurya missile division of the SRF accepts recruits from the surrounding Kirov Oblast into its formations, and has established close ties with regional authorities in order to alleviate some of its more pressing social problems. This cooperation has also resulted in the relaxation of the security regime at the unit by opening it to local commercial interests, in return for their help in maintaining the unit's facilities. Another group that appears to have taken advantage of the more relaxed regime are scrap metal scavengers, who have scoured the wreckage of demolished intermediate-range ballistic missile silos.[44]

It is apparent from the character of such sponsorship arrangements that in most cases they are made for the mutual economic benefit of the military unit and sponsoring entity. In view of the negative reputation that the armed forces have acquired as a result of frequent reports of hazing, the Chechen war, and other problems affecting the military, the concept of sponsorship may have considerable grass-roots support, since it provides some assurance to the families of draftees that their sons will be sent to units that are relatively well off materially and are under a certain degree of external oversight by city representatives. It should be noted that the sponsorship arrangements that include the provision of goods and services, which frequently include the construction of housing for discharged officers in the sponsoring cities, also play an important economic role for the cities by providing an economic stimulus to their local industries. They also represent a locally

---

[42] "Goroda idut na pomoshch podshefnym lodkam," *Vesti*, October 14, 2000, pp. 1, 5.

[43] Aleksandr Kharchenko, ITAR-TASS, June 25, 2003; in "Russian Defense Minister Welcomes Regional Initiative on Military Unit Patronage," FBIS Document CEP20030625000120.

[44] Gennadiy Miranovich and Aleksandr Dolinin, "Yurya is not giving up," *Krasnaya zvezda*, April 16, 2003; in "Russia: status, prospects of RVSN missile division garrison at Yurya," FBIS Document CEP20030416000394.

generated substitute for the defense orders that during the Soviet era were the mainstay of many Russian factories. By providing goods and services for the military directly, rather than indirectly via tax revenues to the federal government, regional governments create a measure of decentralization within the military budget.

Although the military may be forced to accept such limitations on its prerogatives out of sheer economic necessity, sponsorship agreements nevertheless carry some risks. Patronage agreements that specify that military units accept conscripts from sponsoring cities or regions may affect the quality of the conscripts if they bypass the armed services' selection processes, particularly in units that have access to nuclear weapons or reactors. In an SRF unit on the Kamchatka Peninsula, one of its soldiers killed three and wounded one of his colleagues while performing guard duty in January 2000. Both the killer, who was subsequently determined to suffer from psychiatric problems, and his victims were from Petropavlovsk-Kamchatskiy.[45] Although it is not clear whether the servicemen involved in the incident served at the unit thanks to a patronage agreement, the high number of soldiers from that city suggests they were.

Apart from the possibility of bypassing the military's requirements and specifications for its equipment, direct provision of equipment to the military by regional or city governments carries with it the potential for creating considerable procurement chaos. The provision of individual submarines with differing sets of rescue apparatus and other equipment, as proposed by the Omsk government, could considerably complicate crew training, not to mention maintenance procedures. In the long run, such de-standardization of equipment could undermine the safety of submarines.

Therefore, although direct economic assistance under sponsorship arrangements is not an attempt to undermine the military's chain of command—but rather is carried out with the twin goals of aiding the military and furthering the economic interests of the sponsoring entities—such arrangements alter the degree of central control over military assets by limiting the military's freedom of action in the areas of personnel allocation and military procurement. Moreover, such assistance is unpredictable and is allocated unevenly, and is not a substitute for steady and sustained financing of the military by central authorities. Remarkably, Putin's success in tightening control over the regions has not resulted in the elimination of the institution of sponsorship, although it may have resulted in greater eagerness on the part of regional governments to assist military units. Furthermore, Ivanov's encouragement of emulation of the Tver Oblast's efforts to assist SRF units on its territory suggests that—for the foreseeable future—the Russian government will not be able to forgo direct regional economic assistance to military units.

---

[45] Feliks Fokin, "V raketchikov strelyal okkultnykh del master," *Kamchatskoye vremya*, January 20, 2000, p. 1.

Corruption and Organized Crime

A final area of concern is the threat posed to the control of nuclear forces by the increasing power of organized crime. Many criminal groups maintain intimate ties with regional governments and businesses. While Russian organized crime organizations have used the military as a source of conventional weapons and high explosives and have recruited veterans of elite military units, their interest in the military to date has not extended to nuclear weapons. Rather, by infiltrating the military, organized crime groups threaten to undermine combat readiness, and also pose the danger of increasing factionalization within the military, which may lead to resistance to military reform and even insubordination. All of these factors affect the safety and security of Russia's nuclear arsenal. The criminal groups' attempts to establish contacts in the various units have been facilitated by the military's social problems.

The dire financial situation of the Russian military has caused a rapid rise of the in-service crime rate. In some cases, crime has directly threatened the safety and security of Russian nuclear forces. The rate of economic crime in the armed forces has been also increasing in recent years. In 2000, their number doubled in comparison with 1999. The total loss to the MOD in 2000 caused by theft and misappropriation was estimated at 4 billion rubles, 16 times more than in 1999.[46] One of the more prevalent types of crime in Russia is the theft of metals to sell as scrap. Often the "scrap metal" in question consists of electronic components that are valued for the nonferrous and precious metals they contain. In some cases, such "scrap metal" originates from bases controlled by the SRF or other nuclear-capable services. In October 1999, the Northern Fleet brought a criminal case against a group of high-ranking officers who were involved in selling silver extracted from stolen electronic components in torpedoes. This practice was discovered after an investigative commission began to study the causes of torpedo failures during exercises. Another group of officers serving at a submarine unit headquarters illegally sold air purification equipment containing palladium for over $180,000. In one incident, palladium was stolen from a submarine reactor control system.[47] In yet another case, the Vladivostok FSB office confiscated over 3 kg of uranium alloy as a result of an operation against a group involved in selling equipment stolen from naval facilities and defense industry plants in the region.[48]

The majority of economic crimes in the military occur in departments that have direct control over the military's financial or economic assets. Methods used to

---

[46] Vladimir Temnyy, "A chto u vas, rebyata, v galife?" Grani.ru website, February 23, 2001, <http://www.grani.ru/>.

[47] Vadim Vladimirov, "The War Will Be Lightning Fast," *Versiya*, November 7–13, 2000, p. 3; in "Moscow Weekly Looks at Increasing Crime in the Military," FBIS Document CEP20001214000292.

[48] Yana Aminyeva, Grigoriy Maslov, Viktoriya Chernysheva, "Organized Crime of the Far East," Compromat.Ru website, <http://www.kompromat.ru/>, February 27, 2001; in "Organized Crime in Russia's Far East," FBIS Document CEP20010227000314.

defraud the military include taking personal loans in the name of the MOD. Officers also semi-legally lease military bases or buildings to civilian firms. As of 2001, there were 1,509 vacated military bases in Russia, the official value of which was estimated at 30 billion rubles. Only 32 military facilities have been officially sold or transferred to other users. Many of the remainder have changed hands illegally, usually with the collusion of interested officers.[49] According to the Russian Federation's Audit Chamber, which conducted a spot inspection of a number of military organizations, there was a loss of 146 million rubles of government funds and equipment in the Leningrad Military District alone. Fraud took the form of purchasing substandard fuel at premium prices, selling military apartments to civilians, illegally leasing military properties, and selling military equipment that had been reported as destroyed.[50] Even as military training has been curtailed due to fuel shortages, paradoxically, fuel is one of the most commonly stolen commodities in the military. Over 2,200 tons of fuel and other petroleum products were stolen from military bases in the first seven months of 2001 from the Volga and Urals MDs and in the Pacific Fleet alone.[51]

Even the elite SRF and other components of Russia's strategic nuclear arsenal are not immune to economic crime. In December 2000, the Main Military Procurator's Office brought charges against several senior officers at the Plesetsk Cosmodrome. The officers were charged with the theft of nonferrous metals and the misappropriation of funds.[52] Two graduates of the Rostov SRF Institute were apprehended in June 2002 while attempting to sell plastic explosives in the center of Rostov.[53] In January 2003, 10 officers belonging to the 22nd Heavy Bomber Division were arrested on suspicion of embezzling $47,000 from their unit.[54] Even senior officers, with seemingly little financial incentive have been implicated. Major General Karavaytsev, a deputy commander of an SRF unit based near Vladimir, was convicted of abuse of office in 1999. However, before Karavaytsev's crimes were discovered, he was considered to be one of the most outstanding performers in the SRF and his unit received awards for excellent results in training, suggesting either lax supervision or complicity by his superior

---

[49] Vladimir Temnyy, "A chto u vas, rebyata, v galife?" Grani.ru website, February 23, 2001, <http://www.grani.ru/>.

[50] Vladimir Temnyy, "Trofei mirnogo vremeni," Grani.ru website, May 16, 2001, <http://www.grani.ru/>.

[51] "Zaversheno rassledovaniye faktov o kolossalnykh khishcheniyakh v Raketnykh voyskakh strategicheskogo naznacheniya," *Moskovskiy komsomolets*, March 28, 2001, p. 1; in WPS Oborona i Bezopasnost, March 30, 2001; in Integrum Techno, <http://www.integrum.ru/>.

[52] Dmitriy Vladimirov, "Plesetsk Space Center Stolen," *Izvestiya*, December 22, 2000; in "Plesetsk Theft Case May Mean Replacement of 'Entire' Command," FBIS Document CEP20001222000170.

[53] ORT, June 15, 2002; in "Russia: Graduates of military institute caught selling explosives," FBIS Document CEP2002615000018.

[54] Nadezhda Andreyeva, "Odni letayut, drugiye voruyut," *Vremya MN*, January 22, 2003, p. 2; in Integrum Techno, <http://www.integrum.com>.

officers.[55] An investigation of the SRF in March 2001 revealed corruption at high levels of the service. A total of 11 officers were implicated in the case, including a former SRF Equipment and Automation Directorate Chief. This case represented only a small fraction of the corrupt activities in the SRF. An MOD audit conducted at the same time as the investigation allegedly uncovered losses of about $1.8 million worth of SRF equipment.[56]

Organized crime's penetration of Russia's military is facilitated by the fact that many units, including elite SRF divisions, are forced to engage in economic activities in order to sustain themselves. The practice of organizing food production and processing enterprises is still widespread in the Russian military. According to Lieutenant Colonel Sergey Uglov, the SRF food service chief, some 40 percent of the SRF's food needs were met by internal enterprises in 1999. Uglov hoped that the SRF would increase that figure to 60 percent, and that the SRF would expand its food processing facilities.[57] While such practices relieve some of the burden, they create opportunities for regional, possibly criminal, interests to penetrate units, since internal enterprises cannot avoid entering into dealings with outside entities.

In addition to costing the military large sums of money, undermining morale, reducing combat readiness, and increasing the danger of accidents involving nuclear weapons, economic crime has also had the effect of dividing loyalties within the military. As a result of illegal economic activities, large numbers of officers have become closely aligned with regional economic interests (and associated organized crime groups), making them less loyal to the federal government and more likely to resist reforms that may hurt their economic interests. Although this division of loyalties also has the inadvertent positive side effect of reducing the risk of a social explosion in the military and making it more difficult for the military to act in a unified fashion against civilian authority, on balance, it is a negative phenomenon. In some cases, the division of loyalties has already reached the point of passive or active resistance to orders, although fortunately there have been no reported cases of such insubordination affecting nuclear units. To cite one example, when Lieutenant General Mukhamed Batyrov, the commander of ground units in Kamchatka, was ordered in 1998 to transfer his units to the control of the Pacific Fleet, he resisted the order. He also claimed that the commander of the Far East MD, Colonel General Chechevatov, gave him a verbal order to stop equipment transfer. Eventually, Chechevatov convinced the

---

[55] Marina Gridneva, "Amnesty in Uniform," *Moskovskiy komsomolets*, November 30, 2000, p. 2; in "Eight Russian Generals Evade Corruption Charges After Amnesty," FBIS Document CEP20001130000365.

[56] "Zaversheno rassledovaniye faktov o kolossalnykh khishcheniyakh v Raketnykh voyskakh strategicheskogo naznacheniya," *Moskovskiy komsomolets*, March 28, 2001, p. 1; in WPS Oborona i Bezopasnost, March 30, 2001; in Integrum Techno, <http://www.integrum.ru/>.

[57] Vyacheslav Davidenko and Alexander Dolinin, "'Nuclear Missile' Cows, or How to be Food Self-Sufficient for Half a Year," *Krasnaya zvezda*, July 16, 1999; in "RVSN Food Service Chief Interviewed," FBIS Document FTS19990719001010.

central authorities to halt the transfer of troops. It is unclear what motivated the resistance, or the eventual countermanding of orders, but it is likely that commercial factors played a role, since the transfer of units to another command would have meant a loss of control over the financial operations and physical assets of these units.[58]

Military Reform

The influence of regional factors on the federal government's ability to maintain and safeguard its nuclear forces has become the subject of discussion in the context of ongoing military reforms in Russia. In an article published in the May-June 1997 issue of *Voyennaya mysl*, a group of Russian military specialists recommended the creation of mobile reaction forces to defend nuclear reactors and nuclear weapon storage facilities, and transferring nuclear warheads from border areas to more centrally located sites, while taking into consideration the ethnic make-up and the criminal situation in the regions where nuclear weapons were being stored.[59] To preserve nuclear weapon security in this new threat environment, the experts argued, it is of critical importance to prevent commercial and criminal organizations, as well as regional political elites that may be closely tied to organized crime groups, from gaining access to the processes of financing, manufacturing, or storing nuclear weapons. Special care must be taken, they argued, to ensure that units that have access to nuclear weapons are manned only by soldiers with good physical and mental health and no criminal past, and funded in a manner consistent with their elite status and responsibilities. In order to avoid disturbances, nuclear storage facilities should be located in regions where there is a low incidence of ethnic ferment, and organized crime. The team had conducted a threat assessment of each major region of Russia. The Western region was rated as posing a low threat. The situation in the Central region of Russia was also deemed stable, with the exception of areas of the Central regions that border Kazakhstan. The Eastern Siberian region was assessed as being a low threat area, while the Southern region displayed some tendencies toward ethnic conflict. Concerning the nature of the threat facing Russian nuclear bases, the experts identified the possibility of attacks by terrorist groups against missile units in order to spread fear of a possible ecological disaster, the threat of attacks by criminal groups in order to seize weapons and equipment, and local political unrest that may lead to disorder in the immediate vicinity of missile bases.

---

[58] Kimberly Marten Ziak, "Institutional Decline in the Russian Military," in Victoria E. Bonnell and George W. Breslauer, eds., *Russia in the New Century: Stability or Disorder?* (Boulder, Co: Westview Press, 2001).

[59] Ye.M. Suchkevich, V.I. Dumenko, and A.A. Petrov, "Osnovnyye problemy obespecheniya bezopasnosti yadernogo oruzhiya v RF na sovremennom etape," *Voyennaya mysl*, May-June 1997, pp. 15–23.

There are indications that some recommendations of that study are now being implemented. While historically the protection of non-deployed nuclear weapons was split between the 12th Main Directorate (MD) and each service's Sixth Directorate, in a few years the 12th Main Directorate will take over all nuclear storage facilities. During a press conference in October 2001, 12th MD chief General Valynkin said that the consolidation has already started and the directorate would have under its control all warheads and storage facilities belonging to the SRF, Navy, and Air Force within two years. The directorate would not be subject to any cuts, and would place additional emphasis on personnel reliability issues. The directorate has already created mobile reaction units and organized cooperation with MVD and FSB units.[60] As the number of SRF missile divisions decreases, one is also likely to see a greater consolidation of missile units away from areas deemed to possess heightened risks. Significantly, the only SRF regiments to receive the new Topol-M ICBM through the end of 2003 were based in Saratov Oblast, located in the Western portion of Russia, which was deemed to have low threat indicators.[61]

While positive changes in the organizational structures of entities charged with protecting nuclear weapons have already taken place, it is not yet clear what effect reforms underway within the Russian military will have on reducing regional influences. The general trend appears to be toward reducing the share of defense funds spent on strategic armaments. Following a public clash between General Kvashnin, who favored greater emphasis on conventional forces, and former Defense Minister Sergeyev, a ex-SRF commander who favored an emphasis on nuclear forces, President Putin endorsed Kvashnin's reform proposal and replaced Sergeyev with current Defense Minister Ivanov, a former intelligence officer and Security Council Secretary. With the replacement of General Sergeyev, the SRF lost its most influential defender.

The main aspects of the initial reforms under Ivanov and Kvashnin included greater consolidation of Russia's armed services under the MOD's aegis, expansion of the role of the General Staff in military planning and operational issues, enlargement of the powers of MD commanders, elimination of the SRF as a separate branch of service (completed by June 2001), and fairly deep troop cuts. At present, 70 percent of the budget goes for upkeep, and the goal is to reduce the proportion to 50 percent by 2010, with the savings being applied toward training, research, and procurement.[62] According to SRF commander Colonel General Solovtsov, by 2010 the SRF may be reduced to two Army headquarters and 10–12

---

[60] "Minoborony RF: yadernyy terrorizm v Rossii nevozmozhen," Strana.ru website, October 25, 2001, <http://www.strana.ru/>.

[61] Vladimir Muravyev, "No Right to Mistakes," *Armeyskiy sbornik*, December 1, 1999, pp. 20–23; in "RVSN Gen Muravyev: Ground-Based Nuclear Forces," FBIS Document CEP20000112000037.

[62] Dmitriy Trenin, "Military Reform: Can It Get Off the Ground under Putin?" *Demokratizatsiya*, pp. 310-318; "Oblik vooruzhennykh sil silno izmenitsya v budushchem–Sergey Ivanov," Interfax, May 28, 2001.

missile divisions, down from four army headquarters and 19 divisions in the early 1990s. If SRF funding were to be preserved at current levels or increased, these cuts in force structure and personnel would allow some of its pressing needs to be addressed.[63] Moreover, the decision to retain the aged R-36M-series [SS-18 'Satan'] and UR-100UTTKh [SS-19 'Stiletto'] ICBMs, undertaken in the wake of the U.S. withdrawal from the ABM Treaty and subsequent collapse of START II, represents a reversal of the earlier decision to phase out old weapons in favor of the new Topol-M ICBM.[64] This reversal may provide an opportunity to shift funds from the procurement of the single-warhead Topol-M, which lost its *raison d'etre* with the collapse of START II (which banned multi-warhead land-based strategic missiles) to addressing social needs. However, there was no evidence of increased spending in these areas as of early 2004. Other branches of the strategic forces are also going to rely on Soviet-era equipment for the foreseeable future.[65] Since no significant increase in defense spending may be expected in the foreseeable future, the Russian military will continue to face the choice between maximizing the number of deployed warheads by maintaining its Soviet-era delivery systems and addressing its pressing social needs, which have made it vulnerable to crime, theft, and various negative regional influences. The collapse of START II and the adoption of the flexible (if in some areas inadequate) Strategic Offensive Reductions Treaty (SORT, or Moscow Treaty) in May 2002 presented Russia with an opportunity to forgo the expense of reequipping the SRF with single-warhead missiles. The START I counting rules allow a considerable reduction of combat readiness without reflecting negatively on the on-paper Russian strategic nuclear strength, which is still of importance to the Kremlin—in view of its desire not to be relegated to the status of a second-rank nuclear power. It remains to be seen whether that opportunity will be seized, although the apparent decision to forgo further Topol-M production may represent the first step in that direction.

The difficulties experienced by Russia's strategic nuclear forces may be at least partially offset by the gradual assumption of control over all non-deployed nuclear warheads by the 12th Main Directorate, whose status and prestige is not likely to be negatively affected by the reforms, since it is directly subordinate to the General Staff, which is driving the reform. Indeed, it may actually increase. The 12th Main Directorate, by virtue of being a separate, elite, and comparatively well-funded organization that receives extensive assistance from the United States under the Cooperative Threat Reduction program for warhead and delivery system dismantlement, appears to be better shielded from social problems than other components of Russia's nuclear forces. While the 12th Main Directorate is a

---

63 "Zheleznodorozhnyye raketnyye kompleksy ostanutsya v sostave Raketnykh voysk strategicheskogo naznacheniya do 2010 goda," ITAR-TASS, December 16, 2002; in Integrum Techno, <http://www.integrum.ru/>.

64 "Russia to keep multiple warhead nuclear missiles until 2016," Agence France Presse, August 16, 2002; in Lexis-Nexis Academic Universe, <http://www.lexis-nexis.com>.

65 David C. Isby, "Russia to build more SSBNs," *Jane's Missiles and Rockets* online edition, December 1, 2002, <http://jmr.janes.com/>.

heavily professionalized force, in the foreseeable future the Russian military is unlikely to make great strides in the direction of phasing out conscription. In March 2003, Defense Minister Ivanov announced plans to adopt a volunteer manning program, with units belonging to the SRF, Navy, and Air Force transitioning beginning in 2008. However, this plan is likely to be scuttled by a combination of financial difficulties and institutional resistance which, as of early 2004, have prevented the volunteer manning experiment conducted at the Pskov airborne division from becoming a success.[66]

On the positive side, the consolidation of operational command under MOD aegis will likely lead to a greater capacity for joint operations between MOD nuclear asset protection forces and security forces subordinate to other ministries. In some cases local OMON (*Otryady militsii osobogo naznacheniya*, Special Purpose Militia Detachments subordinate to the MVD) units have already participated in mock counter-terrorist operations, including cases involving SRF bases. However, interagency cooperation is hampered by a number of organizational factors.

Conclusion

Although the Russian government is aware of the problems of military regionalism, it appears that this issue will not be fully resolved in the immediate future. The reasons for that are manifold. The Russian political system is still in a state of flux, and the nature of the relationship between the center and the regions is still in the process of negotiation. President Putin's bid to implement "managed democracy" in Russia carries the risk of eliminating the current limited transparency within the Russian military, a development that would hamper efforts to fight pervasive corruption. Even assuming that the political will exists to carry out far-reaching reforms, the Russian government may not be able to afford to implement them. Military reductions and reorganizations usually require considerable short-term expenditures to provide benefits to discharged officers and their families, while the savings from such cuts usually take longer to materialize. The Russian government's inability to provide such additional short-term funding was one of the main reasons that earlier military reform initiatives stalled. There may be a temptation to implement cuts in social spending faster than the cuts in force structure, and the reform's stated purpose of reducing the proportion of personnel-related expenditures to procurement and training-related ones seems to indicate that increasing per capita social expenditures in the military may not be high on the list of reform priorities. The apparent desire to maintain all three legs of the strategic nuclear triad, and to preserve at least the appearance of parity in deployed nuclear warheads with the United States, also diverts resources from

---

[66] Otari Sarkisyan, "Tolko 10,3 protsenta sostoyashchikh na uchete rossiyskikh prizyvnikov nadenut pogony," Regions.ru website, May 6, 2003; in Integrum Techno, <http://www.integrum.com/>.

pressing social needs. Fortunately, the Moscow Treaty may reduce the pressure on the Russian military to retain more strategic systems in service than it can reliably support. Even though the Russian military criticized the treaty for its lack of verification measures and the absence of requirements to eliminate delivery vehicles and warheads, the Moscow Treaty's reduction of strategic warhead ceilings to 1,700–2,200 and allowance of MIRVed ICBMs will reduce the cost of maintaining the strategic nuclear arsenal. Any further mutually agreed reductions of U.S. and Russian strategic arsenals would result in additional savings.

In the long term, reductions in strategic nuclear forces and improvements in the Russian economy may eliminate many of the problems discussed in this chapter. In the short term, however, the most effective policy would be to concentrate control of nuclear warheads in the smallest possible number of units and organizations, and focus on insulating these institutions from the problems that afflict the Russian military as a whole. Doing so would also provide greater assurance that such forces would not be affected by regional influences. Given the difficult financial situation of Russian officers, including those entrusted with the care of nuclear weapons, increasing *per capita* social expenditures would be one of the most cost-effective ways of enhancing the safety of Russia's nuclear weapons. If an across-the-board improvement in the living standards of the Russian military cannot be accomplished in the short run, it would be advisable to concentrate as much responsibility for nuclear weapons in the hands of the fewest possible military organizations (e.g., the 12th Main Directorate), and to ensure the elite status of such organizations. Such measures should also include a reduction in the number of individuals authorized to enter into commercial dealings with civilian entities. Doing away with the practice of maintaining agricultural enterprises, especially in strategic nuclear units, would be another positive measure. Adopting such measures would reduce both the incentive and the opportunity to engage in illicit activities.

It is also clear that military reforms alone will not suffice. The root causes of regional influences on the security and safety of nuclear forces will ultimately be addressed through reconciling the conflicting interests of the Russian federal government and regional authorities. Ubiquitous corruption can be effectively combated only with an open political system and a free media. The establishment of a mutually acceptable and stable division of burdens and responsibilities between the center and the regions would improve the security of nuclear forces by reducing the unpredictability in the interaction between central and regional authorities that now frequently affects Russia's nuclear forces.

PART II
CASE STUDIES OF RUSSIA'S NUCLEAR REGIONS

Chapter 6

Nuclear Issues in the Far Eastern Federal Okrug

Cristina Chuen

Russia's Far Eastern Federal Okrug (district) is unique due to its distance from Moscow, proximity to other nations (including countries of proliferation concern), strong military presence, unstable politics,[1] and the influence of regional actors over local affairs. Although all but one of the nuclear facilities in the district were developed by and for the military—and as such are not technically under local control—regional authorities regularly interact with the facilities and at times play a decisive role in shaping developments at the sites. Citizen groups, especially environmental organizations, also shape developments at local sites, particularly when concern over radioactive materials is involved. The region's economic problems, caused in part by prohibitive transportation and electricity costs, also affect local nuclear installations. Difficulties in securing Russian nuclear materials are all the more dangerous in the Russian Far East, with its nuclear weapons storage sites, nuclear submarine bases, an air base that continues to house nuclear weapon-carrying heavy bombers, and shipyards that construct and repair nuclear submarines, because the region borders North Korea and offers easy access to the sea, a route by which sensitive materials could be transported worldwide.

At present, the Far East is facing a critical period in the development of its nuclear infrastructure as responsibility for decommissioned nuclear submarines is transferred. First, the Ministry of Atomic Energy (Minatom) took over from the Russian Navy, a process that has been ongoing since 1998.[2] Then, on September

---

[1] With the exception of Khabarovsk Kray, all of the regions that make up the Russian Far East have faced political upheavals in the past few years. This is likely to continue for the foreseeable future.

[2] Government Decree No. 518 of May 28, 1998 transferred jurisdiction over everything from defueling and scrapping submarines to storing reactor compartments, spent nuclear fuel, and radioactive waste from the Navy to Minatom. Then, on June 7, 1999, Minatom, the Defense Ministry, and the Ministry of Economics (then still in charge of shipyards—duties now taken over by the Shipbuilding Agency) concluded a three-year plan for the transfer of nuclear submarines subject to dismantlement. The transfer is still under way. Dmitriy Litovkin, "Russia's Naval Doctrine Does Not Consider Problems of Written-Off Nuclear Powered Submarines," *Yadernyy kontrol*, December 15, 1999; in "Russia's Nuclear Sub

15, 2003, a government decree was promulgated according to which Minatom will eventually transfer coordination of dismantlement activities to the Ministry of Economic Development.[3] The transfer of responsibilities between organizations is always difficult, and there are already indications of lapses in security. Any transfers of responsibility increase the tendency of regional actors to play to that ministry, supporting its plans at the local level and in Moscow. The only territory in the Russian Far East that has been unsupportive of Minatom, for instance, is Khabarovsk Kray, which does not have many decommissioned submarines or sites requiring environmental remediation and thus does not require the ministry's financial help.[4]

All of these changes are occurring in an area that already has trouble maintaining its nuclear facilities. In Kamchatka and Amur Oblasts and Primorskiy Kray, frequent blackouts have sometimes included military and even nuclear facilities, in contravention of federal law.[5] Moreover, all of the territories in the Russian Far East face high levels of crime and drug abuse. Economic reforms have hit the region's military facilities particularly hard, at times driving military personnel to supplement their meager incomes through criminal pursuits. As a

---

Recycling Problem," FBIS Document CEP20000322000002, as cited in the NIS Nuclear Profiles Database, <http://cns.miis.edu>.

[3] Charles Digges and Igor Kudrik, "Audit of Minatom reveals millions in misspent cash and lack of control on sub decommission," Bellona Foundation website, December 5, 2003, <http://www.bellona.no>.

[4] There are few decommissioned submarines and no factories receiving dismantlement contracts in Khabarovsk Kray, while any future nuclear power facilities built for the region would chiefly profit the ministry itself and construction facilities located in other regions.

[5] In June 2003, for example, the Primorskiy Kray power provider Dalenergo reduced power supplies to the 13th Pacific Fleet Power Network in Primorye, though not to critical facilities. Earlier, a June 2001 report noted that Dalenergo had cut off Pacific Fleet docking equipment, arsenals and storage facilities, while Khabarovskenergo had cut off military airports, in contravention of Presidential Decree No. 1173 of November 23, 1995. Although Government Decree No. 296 of April 4, 2000 ordered the Ministry of Defense to guarantee financing of heat and electricity supplies, as of November 2000, the Russian Ministry of Defense owed 284 million rubles (about $10 million as of November 28, 2000) to Pacific Fleet electricity suppliers. As a result of problems with Fleet electricity supplies, on November 27, 2000 Russian Prime Minister Mikhail Kasyanov signed Directive No. 1663-r, which established a working commission on the provision of electricity to strategic Pacific Fleet facilities. Nevertheless, as of June 2003, the 13th Pacific Fleet Power Network alone owed Dalenergo 108 million rubles (over $3.5 million). Nikolay Kutenkikh, "Energetiki ne mogut podderzhivat oboronosposobnost strany," *Vladivostok* online edition, <http://vl.vladnews.ru>, January 25, 2002; Viktor Zarembo, "S rubilnikom napereves," *Trud*, June 16, 2001; Vitaliy Denisov, "Zemnyye problemy," *Krasnaya zvezda*, No. 226, November 30, 2000; "Rasporyazheniye pravitelstva Rossiyskoy Federatsii," *Sobraniye zakonodatelstva RF*, No. 49, December 4, 2000, p. 9422; Marina Shatilova, "Primorskiye energetiki vozobnovili ogranicheniya energosnabzheniya obyektov Tikhookeanskogo flota," ITAR-TASS, June 17, 2003.

result, maintaining the safety of nuclear facilities, both active bases and sites being closed down, has become a daunting challenge.

This chapter, based on personal interviews with officials in the region and an extensive reading of local news accounts, examines each of the Far Eastern territories hosting a nuclear installation (see Table 6.1, pp. 107–110), noting how the local political, economic, and criminal situation impacts upon local nuclear sites. While the territories in the Russian Far East share certain difficulties, there are also significant differences between individual territories. Political struggles in Kamchatka Oblasts and Primorskiy Kray have been particularly rancorous. In both territories issues of environmental safety and foreign programs at local nuclear facilities have been political hot potatoes; many local politicians are ready to welcome Minatom initiatives, while others are keener to cooperate with foreign donors. In Khabarovsk Kray, politics have been relatively stable over the past eight years: the governor there brooks no opposition and is standing firm against the construction of any nuclear power plants in his territory. He has also used his strong influence in the territory for local economic gain, even when that means risky military exports. Many politicians in the Far East, particularly those recently elected or appointed, have yet to establish a firm policy on nuclear issues and therefore might willingly be enlisted in helping to solve relevant problems. In order to engage these politicians, however, one must understand how nuclear issues fit into the nexus of local economic and criminal problems, as well as the political map of the Russian Far East.

Table 6.1 Nuclear Facilities in the Russian Far East

Location and Facility Name	Key Assets
Amur Oblast	
Ukrainka Airbase, Ukrainka (Heavy bomber base)	• 21 Bear H16 aircraft (each carrying 16 AS-15 Kent ALCMs or 6 Kickback SRAMs) • 24 Bear H6 aircraft (each carrying 6 AS-15 Kent ALCMs or 6 Kickback SRAMs) (Nuclear warhead total = 480)[6]
Malaya Sazanka Base, Malaya Sazanka (Former central nuclear weapon storage facility)	• Base active until the mid-1990s. Military has left the base, removing nuclear weapons but leaving rocket fuel and other environmental dangers [7]

[6] START I MoU, July 31, 2001.

[7] E. Goncharov, "Bochki ot Minoborony," *Ekologiya*, November 5, 1999, p. 5.

Table 6.1 Nuclear Facilities in the Russian Far East (continued)

Location and Facility Name	Key Assets
Chukotka Autonomous District	
Bilibino Nuclear Heat and Power Plant, Bilibino	• 4 graphite-moderated boiling-water reactors for combined heat and power production[8]
Possible Floating Power Reactor Sites[9]: Pevek, Provideniya, and Egvekinot	• Possible locations for future floating power reactors, similar to the KLT-40 pressurized water reactors used in nuclear icebreakers
Kamchatka Oblast, Vilyuchinsk (near Petropavlovsk)	
Kamchatka Shipyard (Submarine repair shipyard and waste storage)	• 2 service ships (with spent fuel on board) • Radioactive waste storage, Krasheninnikova Bay • SSN/SSGN dismantlement site
Rybachiy Naval Base (Nuclear submarine base)	• 4 active-duty SSBNs (each with 16 SS-N-18 SLBMs; total nuclear warheads = 192) • 12 active-duty attack submarines • 14 decommissioned attack submarines with fuel • 8 defueled, decommissioned attack submarines[10]
Warhead Storage Facility	• Storage of nuclear weapons, presumably for SSBN fleet[11]
Floating Power Reactor Site	• Future location of floating power reactor, similar to the KLT-40 pressurized water reactors used in nuclear icebreakers [12]

[8] "Bilibinskaya ATETs," Institute of Physics and Power Engineering website, <http://www.ippe.rssi.ru/rnpp/bilibino_eng.html>.

[9] A. Kuznetsov, "AES uchitsya plavat," *Atompressa*, January 19, 2000, p. 2.

[10] Mikhail Netecha, presentation on radiation inspections of Far East bases and remediation problems, September 15, 2002, conference on "Ecological Problems in Nuclear-Powered Submarine Dismantlement and the Development of Nuclear Power in the Region," Vladivostok, September 16–20, 2002.

[11] The U.S. Department of Defense publication *Soviet Military Power* reported "nuclear weapons stockpile concentrations" in the Petropavlovsk area. Cited by Joshua Handler, "Lifting the lid on Russia's nuclear weapon storage," *Jane's Intelligence Review*, August 1999, p. 21.

[12] "Minatom Rossii nachinayet finansirovaniye stroitelstva na Kamchatke plavuchey atomnoy elektrostantsii," Strana.ru website, February 21, 2002, <http://www.strana.ru>; A. Kuznetsov, "AES uchitsya plavat."

Table 6.1 Nuclear Facilities in the Russian Far East (continued)

Location and Facility Name	Key Assets
Khabarovsk Kray	
Radon Special Combine, Khabarovsk (Radioactive waste storage facility)	• Radioactive waste storage[13]
Central Nuclear Weapon Storage Facility, Khabarovsk	• Nuclear weapons storage[14]
Amurskiy Zavod, Komsomolsk-na-Amure (Submarine construction plant)	• 2 partially completed SSNs, one of which has a fueled reactor. Russia hopes to lease this SSN to India[15]
Rocket Forces Military Unit 25625, Komsomolsk-na-Amure (Central nuclear weapon storage facility)	• Nuclear weapons storage, possibly closed[16]
Zavety Ilyicha, Sovetskaya Gavan (Nuclear submarine base)	• 3 decommissioned SSNs, with fuel (some of this fuel is damaged). As of April 2001, the SSNs were scheduled to be transported to Zvezda Shipyard in Fall 2001.[17] However, there have not yet been any reports of their having left Zavety Ilyicha
Primorskiy Kray, Rudnaya Pristan (50 km north of Vladimir Bay)	
Possible Floating Power Reactor Site	• Possible future location of floating power reactor, similar to the KLT-40 pressurized water reactors[18] used in nuclear icebreakers

[13] The facility is actually located in the Bolshekhekhtsirskiy nature reserve, 40 kilometers from the city of Khabarovsk. Irina Belova, "Vostrebovany navechno...." *Priamurskiye vedomosti*, May 16, 2001, p. 4, in WPS *Yadernyye materialy*, No. 23, June 15, 2001.

[14] Handler, "Lifting the lid," p. 22.

[15] Vladimir Urban, "'Nerpa' vsplyvet po komande na khindi," *Novyye izvestiya,* January 26, 2002, p. 2; in Universal Database of Russian Newspapers, Eastview Publications, <http://www.eastview.com>; Ivan Safronov, "India pomozhet Ilye Klebanovu zarabotat $4 mlrd," *Kommersant*, June 5, 2001.

[16] The facility is mentioned as operational in Handler, "Lifting the lid," p. 22. However, a February 2001 Khabarovsk media report states that "until recently" a facility near Komsomolsk-na-Amure held 23 nuclear warheads in storage. This would suggest that the warheads are no longer at the site. "'Triumfalnyy' marsh po yadernomu ostrovu," AIF-Dalinform (Khabarovsk), No. 7, February 2001, pp. 1, 4, in WPS *Yadernyye materialy*, March 16, 2001.

[17] Svetlana Komarova, "Ot grekha podalshe," *Priamurskiye novosti*, April 11, 2001.

[18] Kuznetsov, "AES uchitsya plavat."

Table 6.1 Nuclear Facilities in the Russian Far East (continued)

Location and Facility Name	Key Assets
Primorskiy Kray, Shkotovo Peninsula (35 km east of Vladivostok)	
Pavlovsk Bay, Razboynik Bay, and Abrek Cove (Naval bases)	• 1 active-duty SSN, probably to be transferred to Kamchatka Oblast. In October 2000 the Russian General Staff directed the Navy to disband the Pavlovsk Nuclear Submarine Base[19] • 25 decommissioned attack submarines with fuel (of which 3 are contaminated) • 5 defueled, decommissioned attack submarines[20] • 1 nuclear-powered missile cruiser, in reserves[21] • 1 nuclear-powered communications ship, in reserves[22] • Reactor compartment storage
Warhead Storage Facility	• Storage of nuclear weapons, presumably for SSBN fleet[23]
Site 32 and Site 86, Sysoyeva Bay (Submarine waste storage facilities)	• 3 PM-124 class service ships (holding 1,680 spent fuel assemblies; one ship is contaminated, another holds damaged fuel assemblies) • Spent fuel storage (8,400 fuel assemblies) • Highly radioactive waste storage
Chazhma Ship Repair Facility (Submarine repair and refueling facility)	• Fresh and spent fuel storage (over 2,000 kg)
Zvezda Far Eastern Shipyard (START-designated SSBN dismantlement facility)	• Submarine dismantlement facility • Underground liquid radioactive waste storage • *Landysh* floating radioactive waste filtration facility • Interim decommissioned submarine storage

[19] Yevgeniya Lents, "I na Tikhom okeane svoy zakonchili pokhod. Na Dalnem Vostoke rasformirovyvayetsya flotiliya podvodnykh lodok," *Segodnya,* October 13, 2000.

[20] Netecha, presentation on radiation inspections of Far East bases and remediation problems.

[21] Igor Vandenko, "The Cemetery for Healthy Ships: The Pacific Fleet Can Oppose 17 American Missile Cruisers With Only Two. At the Same Time, Entirely Combat Capable Ships Are Rusting While Laid Up," *Novyye izvestiya*, July 15, 1999; in "Cemetery for 'Healthy Ships' in Pacific Flt," FBIS Document FTS19990730000098.

[22] Ibid.

[23] The U.S. Department of Defense publication *Soviet Military Power* reported "nuclear weapons stockpile concentrations" in the Vladivostok area. Cited by Handler, "Lifting the lid."

Chukotka Autonomous District

The Chukotka Autonomous District is home to a nuclear heating and power plant and has been designated by Minatom as a future site for experimental floating nuclear power plants. Chukotka is second in the Far East and third in the country in terms of budget expenditure per capita and is the national leader in terms of the number of loss-making enterprises.[24] The territory relies on nuclear power because all fuel has to be shipped to the district, making electricity and heat extremely expensive. Chukotka's political situation is unusual in that its governor, Roman Abramovich, is one of Russia's "oligarchs," and has had close financial ties to Minatom.[25] Thus, understanding Chukotka's nuclear politics requires both an examination of local relationships (chiefly, the struggle over whether to expand nuclear power or develop alternative power sources) and a look at Abramovich's financial and political relationships in Moscow, which impact upon his activities in the Far East.

Bilibino Nuclear Plant Safety and Possible Future Alternatives

Chukotka continues to rely on the Bilibino Nuclear Heat and Power Plant, completed in December 1976, for much of its electricity and heating needs.[26] Constructed with great difficulty in the Russian Arctic, the Bilibino reactors are surrounded by ordinary building walls, not a radiation containment structure.[27] Bilibino NPP does not meet current Russian safety standards. But the reactor was conservatively designed, with low temperatures and a large heat-sink capability. Safety has been further improved via the International Nuclear Safety Program of the U.S. Department of Energy (DOE).[28] Under this program, DOE has focused on

---

[24] Vladimir Yemelyanenko, "Chukotka Romance," *Izvestiya* online edition, October 31, 2000; in "Roman Abramovich's Chukotka Gubernatorial Electioneering Eyed," FBIS Document CEP20001031000158.

[25] The MDM Group, with which Abramovich is affiliated, was the owner of Konversbank when it handled most Minatom finances. Yuriy Veretennikov, Nikolay Gorelov, Aleksey Grivach, et al., "Sdelki goda," *Vremya novostey*, December 27, 2002, <http://www.vremya.ru>.

[26] A. Gagarinski, "Bilibino," *Nuclear Engineering International*, January 1995, pp. 14–15.

[27] The containment building is a massive structure that surrounds the pressure vessel and the steam generators. Containment structures are commonly made of strongly reinforced concrete, in some cases lined with steel, in others, there is a separate inner steel containment vessel. The containment is intended to retain activity released during accidents and is also believed capable of protecting a reactor against external events such as an airplane crash. David Bodansky, *Nuclear Energy: Principles, Practices, and Prospects* (Woodbury, NY: American Institute of Physics, 1996).

[28] According to Farit Toukhvetov, former director of Bilibino, the plant has a unique cooling system specially designed to cope with arctic conditions, which leave the region short of water that is not frozen. "Its turbine condensers are cooled not by water losing its heat in the usual cooling towers but in a special system of air heat exchangers, making ample use of a

improving the safety of day-to-day operations.[29] Several times Rosenergoatom, the Russian nuclear power operator, has even closed the plant, citing safety reasons.[30] Although the aim of assistance projects was to improve safety until the reactors could be shut down, in August 2003 Minatom decided to extend the service life of Bilibino's first reactor, shut down earlier in the year when its designed service life expired.[31] Current Minatom plans call for the extension of service lives wherever possible. Presidential Representative to the Far East Konstantin Pulikovskiy has stated that closing the NPP is a "road to nowhere." Instead, he wants the government to concentrate on increasing plant profitability.[32] The plant has been caught in a web of triangular debt for years, with periodic infusions of capital from Moscow to keep it running.[33] However, the distance of Bilibino from the populated parts of the country makes improving profitability difficult. It has also made it hard for Russia's nuclear monitoring agency, Gosatomnadzor, to inspect the plant: there have been no inspection trips for the NPP since 1998, for lack of money.[34]

Future Nuclear Construction Plans

A plan to site a new floating nuclear plant in Pevek, Chukotka, outlined in a March 2000 Government Decree, has been postponed. Citing financial and technical issues, Minatom will instead build floating nuclear plants for Severodvinsk, Arkhangelsk Oblast, and Vilyuchinsk, Kamchatka Oblast, first.[35] However, Pevek

---

cheap and plentiful supply of very cold air." The air is pumped through radiators 6 m long and 2.5 m thick by ventilators 3.5 m in diameter. In summer additional coolers are used. Valentin Zazivnov and Farit Toukhvetov, "Bilibino: The world's remotest plant," *Core Issues*, January-May 1999, <http://www.world-nuclear.org/coreissues/1999/no1/focus/index.htm>; Pacific Northwest Laboratory website, <http://insp.pnl.gov:2080/?profiles/bilibino>.

29 For details, see the Pacific Northwest Laboratory website, <http://insp.pnl.gov:2080/?profiles/bilibino>.

30 Anna Bakina, ITAR-TASS, August 20, 1998; in "Russian Company Ready to Propose Closure of Nuclear Station," FBIS-SOV-98-232.

31 Anna Kurbakova, Olga Shipsha, "Chukotka Nuclear Power Station To Resume Full Capacity in Jan 04," ITAR-TASS, August 26, 2003; in FBIS Document CEP20030826000140.

32 To that end, he advocates the creation of a northern energy system, linking the NPP to districts facing power deficits. Energiya Vostoka, "Pulikovskiy ne isklyuchayet sozdaniya yedinoy severnoy energosistemy," in Sakha Net, September 28, 2001, <http://www.sakhanet>.

33 See, for instance, Government Directive No. 1651-r, October 16, 1999, *Sobraniye zakonodatelstva Rossiyskoy Federatsii* 43, October 25, 1999, p. 9818.

34 2002 Gosatomnadzor activity report, February 2003, <http://www.gan.ru>.

35 Oleg Demenin, "Pod flagom 'maloy energetiki'," *Stroitelnaya gazeta,* March 24, 2000; Yevgeniya Borisova, "Russia Plans to Build a Floating Power Plant," *Moscow Times*, March 15, 2001; "Start khorosh. Ne spotknutsya by," *Kurs*, No. 17, December 15, 2000; Yevgeniy Sivayev, "Bespredelno mirnyy atom prigrozil Kamchatke mikrorentgenom," *Kamchatskoye vremya* online edition, October 29, 2001, <http://troyka.iks.ru>.

remains on the list as a site for a future plant, while other potential sites being examined include Provideniya (about 100 km. from Alaska) and Egvekinot, also in Chukotka.[36] The plants, designed by the Afrikantov Experimental Machine Building Design Bureau (OKBM) in Nizhniy Novgorod, are to use KLT-40 pressurized water reactors such as those used in icebreakers, which typically run on highly enriched uranium.[37] Greenpeace is campaigning against the reactors, arguing that they are dangerous and of questionable economic and political value. According to Aleksey Yablokov, a member of the Russian Academy of Sciences, the possibility that the reactors might be leased abroad increases the threat of nuclear terrorism.

Politics and Nuclear Issues

The main political issue in the Chukotka district is electric power.[38] The previous governor of Chukotka, Aleksandr Nazarov, withdrew from the 2000 gubernatorial race after questioning by the Federal Tax Service about electricity subsidies.[39] Abramovich, then Chukotka's Duma deputy, replaced him.[40] Abramovich's 1999 Duma campaign attracted support from then-Minister of Atomic Energy Yevgeniy Adamov, who was then pushing to construct a floating power reactor for Chukotka and has reportedly known Abramovich for quite some time.[41] Anders Åslund, a senior associate at the Carnegie Endowment for International Peace in Moscow, has also suggested that Abramovich has a great deal of influence over Minatom through his financial dealings, although Abramovich denies this influence.[42]

Kamchatka Oblast

Kamchatka Oblast is host to the Pacific Fleet's main naval base, Rybachiy, which homeports the fleet's only four active SSBNs. Kamchatka Shipyard, also known as Shipyard 49K, is also located here. This submarine repair facility is dismantling one SSBN under the U.S. Defense Department's Cooperative Threat Reduction (CTR) program. Both of these facilities are located in the closed city of Vilyuchinsk, a city across Avacha Bay from Petropavlovsk, the oblast capital.

---

[36] Kuznetsov, "AES uchitsya plavat."

[37] Eduard Demin, "OKBM: strategiya razvitiya," *Kurs*, April 7, 2000.

[38] The federal government subsidizes Chukotka's electricity, in order to reduce power rates. As in other regions, however, this practice is not regularized and has led to political fights and accusations of corruption.

[39] *Moskovskiy komsomolets,* October 29, 2000; in "Chukotka Governor Summoned for Questioning by Tax Service on Campaign's Eve," FBIS Document CEP20001030000386.

[40] Aleksandr Minkin, "Kuda maker telyat ne gonyal," *Moskovskiy komsomolets* March 31, 2001.

[41] Roman Shleynov, "Abramovich khochet smenit 'semyu'?" *Novaya gazeta,* April 2, 2001.

[42] Peter Baker, "An Unlikely Savior on the Tundra: A Russian Tycoon Adopts Abandoned Arctic Region, but Why?" *Washington Post*, March 2, 2001.

Kamchatka is also home to a nuclear warhead storage facility, located in the vicinity of the Rybachiy Naval Base.[43]

City-Navy Relations

Vilyuchinsk is a closed city with a popularly elected civilian government. During the past several years, responsibility for the upkeep of housing, streets, and other such basics has been transferred from the Navy to the city. This has sometimes resulted in conflict between the Navy and the city, as the city does not have the money to maintain the facilities and does not spend the money it has according to the desires of the Navy. During the Soviet era, Rybachiy received priority funding. Today, money is sent according to the number of inhabitants in a district. Thus, the part of Vilyuchinsk where shipyard workers live has received increased funding, while the Rybachiy district cannot afford to maintain even basic upkeep of housing and roads. In January 2000, this problem worsened when amendments to Russian tax laws transferred some of the taxes collected in closed cities to the territorial budget (this money used to remain in city coffers).[44]

On the more cooperative side, the city of Vilyuchinsk sponsors the Oscar II-class nuclear-powered cruise-missile submarine (SSGN) *Vilyuchinsk*, based at Rybachiy, providing clothing, food, vacations, and other benefits to the sailors. The city of Petropavlovsk similarly sponsors the submarine *Petropavlovsk-Kamchatskiy* due to the sharp decline in naval funding for the basic needs of sailors. However, the level of sponsorship that these two relatively poor cities can provide is less than the assistance provided by Irkutsk, Krasnoyarsk, Chelyabinsk, Omsk, Tomsk, Podolsk, and Zelenograd, other sponsors of active-duty Rybachiy submarines. Some of these richer cities have suggested that the entire town of Vilyuchinsk ought to be sponsored, but the plan was abandoned due to cost.[45]

---

[43] The U.S. Department of Defense publication *Soviet Military Power* reported "nuclear weapons stockpile concentrations" in the Petropavlovsk area. Cited by Handler, "Lifting the lid."

[44] Formerly, about half of the taxes collected locally were sent to the federal government while the remainder stayed in Vilyuchinsk. After January 1, 2000, approximately half were sent to Moscow, one quarter to Petropavlovsk, and one quarter remained in Vilyuchinsk. Author's interview with Vilyuchinsk treasury official, June 16, 2000.

[45] On October 18, 2000, representatives of nine Russian cities (Vilyuchinsk, Petropavlovsk-Kamchatskiy, Irkutsk, Krasnoyarsk, Tomsk, Omsk, Chelyabinsk, Podolsk, and Zelenograd) met in Petropavlovsk-Kamchatskiy to form a "sponsorship council" to help maintain the Pacific Fleet nuclear submarine unit based at Rybachiy. The council consists of the cities' mayors. Under the sponsorship agreement, signed at the meeting, the cities will assist sailors in solving housing problems and in finding jobs for those leaving the service. They also pledged to send only the best conscripts from the regions to serve on the submarines. Under the arrangement between the regional authorities and the fleet command this cooperation will last for 55 years. "Soglasheniye o razvitii shefskikh svyazey i sotrudnichestva mezhdu gorodami Rossii, Krasnoznamennoy Operativnoy eskadroy Tikhookeanskogo flota i ee korablyami," *Vesti*, October 14, 2000, p. 5; Nikolay Litkovets, Yuriy Rossolov, "I stala

Meanwhile, several active vessels at Rybachiy have yet to find sponsors, and decommissioned submarine crews have no sponsors at all. The wherewithal of even the wealthier sponsors is limited. In November 2000, the crew of the *Krasnoyarsk* SSGN requested that the Krasnoyarsk governor help them obtain funding for submarine renovations.[46] However, in early 2001, the *Krasnoyarsk* was decommissioned and its weapons removed. In January 2002, the crew again requested assistance, in the hopes that the vessel might be recommissioned. But repairs would have cost an estimated two billion rubles (over $65 million), nearly one-third of the Krasnoyarsk city budget.[47] The practice of sponsorship, while helping to maintain submarines and their crews in the short run, may prove dangerous in the long run: the Navy may increase its demands on sponsors, treating the submarines and their crews as hostages, or the territories may become unable to maintain their sponsorship—on which the Navy relies to maintain its vessels.

Despite the Vilyuchinsk administration's concerns regarding the environment, it has come out in favor of basing a floating power reactor in the bay to provide heat and power to the town.[48] Any dangers a floating power reactor might bring are overshadowed by the dangers of the regular power outages faced by the territory. On January 26, 2002, Kamchatka Oblast's power provider cut off the Russian Military Space Forces Telemetry Center, endangering the operation of military reconnaissance satellites and the international space station.[49] The Rybachiy Submarine Base has also had its power cut off.[50] On November 6, 2002, territorial electricity provider Kamchatskenergo Director Yuriy Delnov complained that the military owed it over 200 million rubles.[51] In the winter of 2003-2004, heating and power provision again ran into difficulties.

---

blizhe Kamchatka..." *Krasnaya zvezda*, October 26, 2000; "Predstaviteli devyati rossiyskikh gorodov sozdali sovet shefov soyedineniya atomnykh podvodnykh lodok Tikhookeanskogo flota," ITAR-TASS, October 18, 2000.

[46] "Ekipazh atomnoy podlodki 'Krasnoyarsk' poprosil Aleksandra Lebedya pomoch v finansirovanii remonta submariny," ITAR-TASS, November 17, 2000.

[47] Yevgeniy Latyshev, "Sukhoputnyye moryaki," *Novyye izvestiya*, January 22, 2002.

[48] "Na Kamchatke dan khod proyektu stroitelstva plavuchey atomnoy elektrostantsii" and "Proyekt stroitelstva plavuchey atomnoy elektrostantsii u beregov Kamchatki odobren administratiyey oblasti."

[49] Oleg Zhunusov and Dmitriy Safonov, "Otklyucheniya energii dostigli kosmosa," *Izvestiya*, January 27, 2002, <http://www.izvestia.ru>.

[50] In July 2000, fleet commanders had electricity only a few hours per day, data transmission equipment was down nine hours per day and submarine crews were reduced to preparing meals over wood fires. Agence France Presse, July 26, 2000, in "Power Cuts Put Control of Russian Pacific Fleet at Risk," FBIS Document EUP20000726000139.

[51] Anna Gromova, "Zima—nash ekzamenator," *Novaya kamchatskaya pravda*, November 6, 2002.

Economic Issues and Crime

One of the poorest territories in Russia, Kamchatka Oblast's economic woes have a clear effect on territorial military installations. The territory is less able to subsidize local military units than are Russia's richer territories. While officers and men in Kamchatka Oblast take jobs off base to supplement their meager income, these jobs are scarcer than in some other territories. The high level of criminality in Kamchatka Oblast is a further temptation, making it easy for servicemen to find purchasers for materials stolen from military bases. At the other end of the social strata, criminals have entered into political office.[52] In November 2001, Aleksandr Drozdov, deputy presidential representative to the Russian Far East, charged that more than 50 candidates in legislative races in Kamchatka Oblast and Khabarovsk and Primorskiy Krays have links to organized crime.[53]

As for criminal activity in Vilyuchinsk, crime involving the Kamchatka Shipyard and shipyard workers is low, as they are generally well paid and paid on time. The Rybachiy base is another matter. For instance, in January 2000, sailors bribed a guard and broke into a decommissioned nuclear submarine with two functioning nuclear reactors—but were later caught. They had removed a number of elements from the reactor control unit, nine pipes (worth $3,500 each), and radioactive calibrating metal. Fortunately, they were unable to raise the reactor control rods because a mechanic, acting on his own initiative, had bolted down the central lever the previous day. Had they been able to do so, they might have caused the reactor to overheat and possibly created a dangerous criticality incident, venting radiation and killing perhaps hundreds of people.[54] In another incident, in April 2001, two Navy officers were arrested in Petropavlovsk while trying to sell parts of a submarine's radio-navigational equipment containing radioactive materials. According to Petropavlovsk garrison military prosecutor Yuriy Sazonov, the stolen items could have come from a nuclear submarine. He also said that it was the fifth case involving attempts by military staff to sell military property in 2001.[55] In fact, during a 2001 investigation of thefts of platinum catalysts from

---

[52] In March 2000, Kamchatka Deputy Governor Aleksey Kotlyar was murdered in his home by a hired thug. According to press reports, another Kamchatka politician, Yuriy Babak, the deputy mayor of Yelizovo (a municipality adjacent to Petropavlovsk), hired the murderer. It was Babak's first arrest, but press reports link him to other illegal activities in the past. Vyacheslav Skalatskiy and Svetlana Soloveva, "Samoye gromkoye ubiystvo na Kamchatke raskryto?" *Kamchatskoye vremya*, June 7, 2001.

[53] RIA Novosti, November 30, 2001, as cited in "Nazdratenko Returns to Stump for Candidates," RFE/RL *Russian Federation Report*, Vol. 3, No. 34, December 5, 2001.

[54] Mikhail Druzhinin, "Matrosy razdolbali yadernyy reaktor," *Kommersant*, January 29, 2000; RIA-Novosti, January 31, 1999; Tatyana Oshchepkova, "Na Kamchatke obokrali atomnuyu submarinu," *Yezhednevnyye novosti*, February 1, 2000; Nonna Chernyakova, "Military Admits Theft of Radioactive Metal from Sub," *Vladivostok News*, February 4, 2000.

[55] "Novosti stran SNG," ITAR-TASS, April 17, 2001; Aleksandr Arkhipov, "Na Kamchatke zaderzhany dva ofitsera-moryaka, pytavshiyesya torgovat radioaktivnym tovarom," ITAR-

submarine oxygen-generating respirator canisters, inventory checks found that platinum catalysts had been stolen from most Rybachiy submarines.[56]

While the thieves in the incidents related above did not appear to have buyers in mind, this is not always the case. In February 2001, three military men stole instruments from Mi-8 helicopters that are used to measure ice-cover. The expensive equipment, which contains strontium-90 (at levels of 25 microRoentgens per hour one meter from the equipment), was destined for a Chinese buyer. Officers of the Kamchatka Oblast Counter-Organized Crime Department thwarted the sale.[57]

While reported thefts of radioactive materials are few, the general level of theft on the base is quite high. In a June 14, 2000 interview, a former submarine commander who continues to work in Vilyuchinsk stated that criminal organizations in Kamchatka Oblast regularly provide Rybachiy recruits with cards listing prices and diagrams of what metals to steal from submarines. He believes there are quite a few successful thefts, as submarines are often found to be missing equipment.[58] The high level of systematization these metal thefts involve suggests that should a buyer for nuclear equipment come forward, criminal organizations may well be successful in stealing it. Financial crimes also endanger the safety of the Rybachiy base. In 2000, examinations of the finances of local military bases determined that the Radiation, Chemical and Biological Protection Service had stolen or lost 4.5 million rubles (nearly $160,000) between 1998 and 1999.[59]

Another factor exacerbating safety problems in Kamchatka Oblast, as elsewhere in the Russian Far East, is the high rate of drug abuse. Narcotics abuse is a particular problem in the military, and is likely to affect guard units as well as others safeguarding radioactive materials. In November 2001, Presidential Representative Pulikovskiy created a special interdepartmental commission to

---

TASS, April 23, 2001; Aleksandr Arkhipov, "Protiv dvukh ofitserov, zaderzhannykh na Kamchatke po podozreniyu v torgovle radioaktivnymi priborami, vozbuzhdeno ugolovnoye delo," ITAR-TASS, April 28, 2001; "Ofitsery VMF torgovali komponentami yadernogo reaktora," *Vremya MN*, April 24, 2001.

[56] Svetlana Oskina, "Zhazhda platiny sgubila," *Kamchatskoye vremya*, February 22, 2001.

[57] Yevgeniy Pozdnyakov, "Voyennyye pribory, soderzhshchiye radioaktivnyye veshchestva, pytalis nezakonno prodat v Kitay," NTV television, February 16, 2001.

[58] Author's interview with former Russian submarine commander, Petropavlovsk, June 14, 2000.

[59] Officials found responsible for financial crimes included the head of the inspectorate's section charged with monitoring financial activities, the head of the division's financial service, and several unit commanders. "Rezultaty finansovoy inspektsii voysk kamchatskoy gruppirovki dobavyat raboty prokurature i sudu," Agenstvo voyennykh novostey, February 9, 2000; "Po resultatam finansovoy proverki v kamchatskoy gruppirovke nakazano bolee 20 starshikh ofitserov," Agenstvo voyennykh novostey, March 23, 2000.

study and fight the district's illegal drugs, noting that drug-related crimes have increased each year.[60]

Kamchatka Oblast Politics and Nuclear Issues

Kamchatka politicians have a great deal of influence over the nuclear installations on the peninsula. Although Governor Mikhail Mashkovtsev[61] has never publicly stated his policies regarding local naval facilities and nuclear issues,[62] since his January 2001 election he has fostered good relations with Minatom, and has consolidated political power on the peninsula.

The largest problem facing Kamchatka Oblast is electricity provision. Kamchatka energy producer Kamchatskenergo is against the introduction of nuclear stations into the oblast, pointing out that the power crisis is caused mainly by payments crisis, causing local heat and power stations to operate only at 30 percent capacity.[63] The governor, nevertheless, has come out in support of a floating nuclear power reactor, in large part because funding for the plant will reportedly come from Moscow.[64]

Additionally, territorial politicians periodically have argued about the desirability of reconfiguring nuclear submarines to provide electricity to

---

[60] Maksim Gladkiy, "Udobnaya doroga. Dalniy Vostok stal zonoy narkotrafika," *Vremya novostey*, November 30, 2001, p. 2; in Universal Database of Russian Newspapers, Eastview Publications, <http://www.eastview.com>.

[61] Mashkovtsev, head of the local Communist Part branch, replaced long-time Governor Vladimir Biryukov in a very dirty campaign in which Kamchatka's State Duma deputy, Vice-Admiral Valeriy Dorogin, challenged Biryukov's protégé, First Deputy Governor Boris Sinchenko. Mashkovtsev, a dark horse, narrowly won the vote. His election led to months of protests by angry citizens. First seen as a weak governor, who might be co-opted by Biryukov, he has since succeeded in strengthening his independent position.

[62] Mashkovtsev had made anti-American comments, blaming the C.I.A. for the existence of United Nations-designated natural parks in Kamchatka (due to their protected status, minerals cannot be mined in the parks), for instance. However, it would appear that these remarks are chiefly aimed against environmentalists, foreign and domestic, not the United States more generally or possible U.S. assistance for submarine dismantlement. Yevgeniy Sivayev, "Eldorado po-kamchatski," *Kamchatskoye vremya*, January 25, 2001, p. 13.

[63] K. Marenin, "Atomnyy flot—na voynu s energokrizisom," *Rybak Kamchatki*, February 15, 2001, p. 8.

[64] According to Vilyuchinsk Mayor Aleksandr Markman, the 70-megawatt floating reactor project is part of the Russian program for the development of nuclear energy in 2000-05, and would be located in Avacha Bay. Estimates of floating reactor construction costs vary widely; the Vilyuchinsk plant is estimated to cost over $200 million, money which will be raised by the Ministry of Atomic Energy. "Na Kamchatke dan khod proyektu stroitelstva plavuchey atomnoy elektrostantsii," Vostok-Media, September 7, 2001; "Proyekt stroitelstva plavuchey atomnoy elektrostantsii u beregov Kamchatki odobren administratiyey oblasti," ITAR-TASS, September 7, 2001; Yevgeniy Sivayev, "Bespredelno mirniy atom prigrozil Kamchatke mikrorentgenom," *Kamchatskoye vremya*, October 29, 2001.

Kamchatka Oblast's cities.[65] Several times during power outages in the past few years, nuclear submarine reactors have indeed been engaged so that power could be fed to the naval base.[66] However, the amount of electricity generated is small, and, according to one former local submarine commander, running the reactors near shore can only be done safely for several days, due to potential coolant problems.[67] Reconfiguring the submarine reactors to provide more electricity (and not propulsion) is an expensive process. One study in 1994 determined that reconfiguring one reactor would cost about 10 billion rubles and take a full year. A later study of converting the nuclear-powered communications vessel *Ural* to electricity production found conversion would cost $65-80 million. Both projects, therefore, were dropped.[68]

Kamchatka politicians also often have argued about the dismantlement of Rybachiy submarines. Kamchatka Oblast's State Duma deputy (and former head of the Northeastern Group of Troops and Forces) Vice-Admiral Valeriy Dorogin has spoken out against the dismantlement of submarines on Kamchatka Oblast as well as the local storage of spent nuclear fuel and radioactive waste. He has alleged (falsely) that the United States continues to dump liquid radioactive waste in the oceans, and suggested that it would be safer for Kamchatka Oblast to do the same. He has also suggested scuttling submarines in the ocean, rather than dismantling them.[69] Finally, he has at times asserted that U.S. dismantlement assistance programs in Kamchatka Oblast would involve importing American radioactive waste to the region and divulging Russian military secrets. On the other hand, he has been quoted as saying that the United States was welcome to provide its money to dismantle Russian submarines.

The local public has periodically worried about radioactivity emanating from the Kamchatka Shipyard, since news reports in 1991 began to chronicle liquid radioactive waste leaks at Vilyuchinsk. In an effort to improve the situation, the Kamchatka Oblast parliament and the Kedr political party have directly assisted the shipyard by providing funding and equipment, including computers for research on handling radioactive waste.[70] The local citizenry is still very wary of official secrecy regarding the naval base. In November 2000, the media criticized Governor Dorogin for secrecy in his handling of the sinking of two

---

[65] Mariya Shustrova, "Kantemir Karamzin stal kostyu v gorle provorovavsheysya vlasti," *Kamchatskoye vremya*, May 4, 2000, p. 8.

[66] "U ministerstva oborony est i drugiye dolgi, krome voinskikh," *Kamchatskoye vremya*, May 25, 2000, p. 4; Yevgeniy Ustinov, "Korabli, dayushchiye teplo," *Krasnaya zvezda*, May 12, 1999; "Atomnyye podvodnyye lodki tikhookeanskogo flota snabzhayut elektroenergiyey kamchatskiy gorod," *Parlamentskaya gazeta,* November 12, 1998, p. 2.

[67] Author's interview with former Rybachiy submarine commander, Petropavlovsk, June 14, 2000.

[68] Marenin, "Atomnyy flot—na voynu s energokrizisom"; "'Ural' elektrichestva ne dast," *Kamchatkoye vremya*, November 30, 2000, p. 4.

[69] Dmitriy Spitsyn, "Nuzhna li Kamchatke yadernaya svalka?" *Gorodskaya gazeta*, April 20, 2000, p. 2.

[70] "Vse khorosho, prekrasnaya markiza," *Gorodskaya gazeta*, January 18, 2001.

decommissioned nuclear submarines at their piers during 2000.[71] Indeed, protection of radioactive waste stored on the Kamchatka Peninsula appears to be deteriorating. In October 2001, a group monitoring radiation levels in Vilyuchinsk reported that during a two-hour period they observed no guards posted around a radioactive waste storage facility that had formerly been under 24-hour armed guard when the Russian Navy was in charge.[72] In addition, the barbed wire surrounding the facility had large holes that a person could easily pass through. While the Navy was able to order sailors to stand guard at the site, Minatom must pay for guard personnel. Such funding appears to be in short supply.

Khabarovsk Kray

Khabarovsk Kray is home to Amurskiy Zavod (a shipyard that constructs nuclear submarines), the Radon Special Combine radioactive waste storage facility, two nuclear warhead storage facilities (one of which is likely now closed), and a small naval base at Zavety Ilyicha. Unlike some regional leaders, Khabarovsk Governor Viktor Ishayev is a strong national figure and holds great sway over the facilities in his territory. Reportedly, local military leaders consult with him regarding the scheduling of military maneuvers in the territory.[73] Since Moscow remains unable to fully finance military industries and military bases, they depend upon assistance from the territories. This appears to be the source of regional leaders' strong influence over their activities, an influence particularly felt in the Khabarovsk region.

Crime and the Military

While the Khabarovsk Kray economy is somewhat more stable than other Far Eastern economies, it is nevertheless depressed. This has resulted in high crime rates and difficulties maintaining security at local military factories and military bases. For example, the commander of one impoverished Khabarovsk Kray military facility has allegedly allowed Chinese citizens to buy up apartments *inside* the base.[74]

---

[71] The sinking of two decommissioned submarines in Kamchatka has been confirmed by Anatoli Diakov, Director of the Center for Arms Control, Energy, and Environmental Studies, Moscow Institute of Physics and Technology. Anatoli Diakov, remarks at the Russian American Nuclear Security Advisory Council Congressional Strategic Stability and Security Seminar Series, Seminar 2: Cooperative Nuclear Threat Reduction Activities in Russia, Washington, D.C., May 18, 2001; Valeriy Saykov, "Armiya i Dorogin: po raznyye storony barrikady."

[72] Kolesnikov, "Lodki otstoya rzhaveyut i tonut, 'mogilniki'—ne okhranyayutsya, mestnyye vlasti zhdut ot Moskvy ocherednoy federalnoy programmy."

[73] Interview by Monterey Institute researcher with Khabarovsk Kray customs official, December 1999.

[74] Ibid.

To date, most thefts involve metals and spare parts. There have reportedly been attempts by Chinese operatives to persuade workers at Amurskiy Zavod to steal items from the plant.[75] In a case that jeopardized nuclear weapons security, a criminal group ran a used car dealership at Rocket Forces Military Unit 25625 (located near Komsomolsk-na-Amure) in 1998, before the nuclear warheads had been removed from this facility.[76] Despite warning signs that guards would shoot on sight and the mandate for a high level of physical protection at the facility, a journalist observed drivers and purchasers easily passing by the guards at the main gate.[77] The same facility continued to face criminal problems in subsequent years. In early 2000, the base rail service head was charged with stealing and selling two railcars for the transport of special materials, which could be used to transport items without civilian oversight.[78]

As is the case elsewhere in the Russian Far East, the greatest deterrent to stealing radioactive materials is the lack of buyers. However, highly organized criminal organizations suggest that should this situation change, disaster could quickly result. If media reports suggesting that some central nuclear weapons storage facilities have closed are correct, there is one less site needing physical protection in Khabarovsk Kray. Other sites housing radioactive materials remain targets of local criminal organizations.

The State of the Zavety Ilyicha Naval Base

A home port for nuclear attack submarines beginning in 1982, all submarines at the Zavety Ilyicha Naval Base were decommissioned in 1990.[79] Local residents immediately protested against government plans to offload nuclear fuel from the decommissioned submarines in the bay and store the floating hulls at the base. The

---

[75] One of the more alarming thefts at a Khabarovsk military unit involved the smuggling of Su-27 aircraft parts from the Komsomolsk-na-Amure Aviation Production Association and active military units. One of the air regiments had to disband due to a lack of equipment after the thefts. An attempt to steal and sell a top-secret radar complex for the Su-27 aircraft to China, on the other hand, was foiled. ITAR-TASS, February 5, 1999; *Profil*, June 28, 2000.

[76] A February 2001 Khabarovsk Kray media report states that "until recently" a facility near Komsomolsk-na-Amure held 23 nuclear warheads in storage. This would suggest that the warheads are no longer at the site. See "'Triumfalnyy' marsh po yadernomu ostrovu."

[77] Boris Reznik, "'Krysha' dlya sekretnogo yadernogo obekta," *Izvestiya*, January 21, 1998.

[78] Instead of a demotion, the perpetrator was transferred to Moscow. Tatyana Sanina, "Zheleznaya doroga prichastna k khishcheniyam v voyennom garnizone?" *Amurskiy meridian*, March 1, 2001; Maks Molotov, "Prikazano: vyzhit," *Molodoy dalnevostochnik*, July 11, 2001.

[79] Joshua Handler, "Russia's Pacific Fleet: Submarine Bases and Facilities," *Jane's Intelligence Review*, April 1994, p. 170; report by V.A. Danilian and V.L. Vysotsky, cited in Joshua Handler, "The Russian Naval Nuclear Complex," in Busmann, Meier, and Nassauer, eds., *The Nuclear Legacy of the Former Soviet Union: Implications for Security and Ecology*, BITS Research Report 97.1, November 1997, p. 33.

Pacific Fleet responded by stating that it would remove one nuclear submarine from Zavety Ilyicha per year beginning in 1991. While one was removed in 1993, three remained at the base.[80] According to one researcher, seven kilograms of spent submarine fuel (which contains highly enriched uranium) was stolen from Zavety Ilyicha in January 1996.[81] There are no further reports of thefts from the base, but information on activities there is scant.

In April 2001, the Russian Navy again announced that the last three nuclear submarines at Zavety Ilyicha would be transported to the Zvezda Shipyard in Primorskiy Kray for dismantlement. The vessels were the cause of growing local nuclear safety concerns, especially after the Navy discontinued regular underwater checks for damage and corrosion. None of the three has been defueled, and some of this fuel is apparently damaged.[82] As of December 2003, no schedule for the removal of these vessels had been publicly announced.

Khabarovsk's Nuclear Industry

Two Akula II-class nuclear submarines have been under construction at the Amurskiy Zavod shipbuilding plant in Komsomolsk-na-Amure for over a decade and a half. During then-Prime Minister Vladimir Putin's October 1999 visit to the facility, he announced that one submarine would be completed, and the other used for spare parts.[83] However, the latter submarine remains at the yard; construction could resume at some future time.

The former submarine, *Nerpa*, is approximately 80 percent complete, with its reactor fully loaded with nuclear fuel. Its hull was completed in June 2000.[84] However, according to Ishayev, the shipyard was only allocated 5 million rubles in the 2001 budget, while the cost of maintaining the two unfinished nuclear submarines at the facility in their current state was 70 to 80 million rubles ($2.5-2.8 million) per year.[85] Despite some commercial contracts at the plant, division head Yevgeniy Krysyuk noted that in 2000 some plant workers had fainted from hunger, because they were so underpaid that they could not feed themselves

---

[80] Handler, "Russia's Pacific Fleet: Submarine Bases and Facilities."

[81] Rensselaer W. Lee III, "Smuggling Update," *Bulletin of the Atomic Scientists* 53 (May-June 1997), pp. 52–56; Rensselaer W. Lee III, *Smuggling Armageddon: The Nuclear Black Market in the Former Soviet Union and Europe* (New York: St. Martin's Press, 1998), p. 119.

[82] Komarova; "Tri atomnykh podvodnykh lodki Sakhalinskoy flotilii Tikhookeanskogo flota perepravyat v Primorye dlya utilizatsii," ITAR-TASS, April 18, 2001; "Na zavode 'Zvezda' v Primorye budut utilizirovany tri atomnykh podvodnykh lodki Tikhookeanskogo flota," *Finmarket*, April 18, 2001.

[83] Vremya newscast, October 27, 1999.

[84] "APL tretyego pokoleniya 'Bars'," UNIAN, June 26–July 2, 2000; Yevgeniya Lents, "Vozobnovilos stroitelstvo noveyshikh atomnykh submarin," *Segodnya*, June 8, 2000.

[85] Interfax, "Predpriyatiya VPK Khabarovskogo kraya ispytyvayut defitsit v oboronnom zakaze," February 24, 2000.

properly.[86] Interviews suggest that so many key personnel have left the facility because of wage arrears that completion of either submarine might be difficult even if the Ministry of Defense provided the funding.[87]

The shipyard has therefore been seeking new sources of funding to maintain and complete the SSNs. Most notably, they hope to lease the *Nerpa* to India. In January 2002, an official Russian announcement stated that Russia would lease the submarine to India.[88] Amurskiy Zavod and the Khabarovsk Kray government were reported to be strongly behind a lease or sale, while the Russian Foreign Ministry was evidently more wary.[89] Indian reports continue to suggest that negotiations are moving ahead, though recent Russian pronouncements suggest the deal might be off.[90]

Clearly, the Amurskiy plant needs money to maintain, complete, or dismantle the two submarines, as well as to maintain physical protection around the vessels. The Russian government has not provided sufficient funding for any of these tasks, leaving the fueled reactors vulnerable to sabotage. Governor Ishayev has suggested that the territory's frequent electricity cut-offs could also cause a disaster: in June 2001 he was quoted as saying that if a lack of heat causes the reactor to freeze, a "not very small Chernobyl" may result.[91] There do not appear to be any foreign nations undertaking to assist Amurskiy Zavod with protection measures. Until the prospect of a lease to India became likely, the shipyard would probably have welcomed offers to help dismantle both *Nerpa* and its sister ship, instead of completing them, making the full conversion of Amurskiy Zavod possible, so that no future nuclear submarines are constructed at the facility. Indeed, in 1992 President Yeltsin had ordered that nuclear submarine work at Amurskiy cease. However, in addition to the probable lease, the shipyard has recently received its first orders from the Russian Navy in over a decade: in October 2002, the Pacific Fleet contracted it to overhaul nuclear and diesel submarines that had been built at

---

86 Golts, "Ne gluboko eshelonirovannaya oboronka."

87 James Clay Moltz, "Trip Report: Vladivostok and Khabarovsk, Russia," October 15–22, 1999.

88 Ivan Safronov, "Indiya pomozhet Ilye Klebanovu zarabotat $4 mlrd," *Kommersant*, June 5, 2001.

89 Aleksandr Yemelyanenkov, "Zastryavshiy na stapelyakh," *Rossiyskaya gazeta*, December 1, 2001.

90 The Indian press continues to write of a "semi-covert arrangement," while Russian Defense Minister Sergey Ivanov stated that Russia does not plan to conclude a contract for the submarines "in the near future." Manoj Joshi, "Nuclear Wishlist is Heart of Carrier Deal," *Times of India*, January 22, 2004; Interfax, January 19, 2004.

91 "Esli energetiki za dolgi otklyuchat elektroenergiyu, esli ne budet tepla i zamerznet voda v reaktore [strana i region poluchat novyy i] ne ochen malenkiy Chernobyl." As cited in Vasiliy Buslayev, "Kogda i za chto osvobodyat Pulikovskogo?" *Yezhednevnyye novosti*, June 20, 2001.

the plant.[92] In November 2003, the first nuclear submarine arrived at the shipyard for repairs.[93] Thus, the opportunity to convert the yard seems to have been lost.

In addition to Amurskiy Zavod, Khabarovsk Kray is also home to the Radon Special Combine radioactive waste storage facility and a site at which Minatom plans to construct a new nuclear power plant. The Radon plant is located approximately 40 kilometers from the city of Khabarovsk, in the Bolshekhekhtsirskiy nature reserve.[94] Built for the disposal of radioactive waste from medical, scientific, and technical facilities, Radon is not subordinate to Minatom and does not handle waste from nuclear power plants, although it does take solid radioactive waste from Amurskiy Zavod and the *Landysh* liquid radioactive waste processing facility in Primorskiy Kray. On May 16, 2001, *Priamurskiye vedomosti* reported that the facility was to receive an alarm system during the course of 2001. A fence and a 10-meter control zone already surrounded the facility.[95] In 2002, new radioactive waste transferring equipment was installed, to move existing waste into safer storage and to improve handling of new waste.[96] Radon's location, in a protected forest, has led some to worry about possible environmental contamination. Conversely, the forest has at times threatened the safety of the facility; in October 2001, a major forest fire threatened Radon.[97] However, no plans to further update protection at the facility have been made public.

Khabarovsk Politics

Governor Ishayev's relationship to nuclear issues is not clear-cut. His administration has come out against the construction of any nuclear power reactors in the territory, regardless of Minatom wishes. Ishayev also spoke out against Minatom plans to import spent nuclear fuel, likely due to the transit issue.[98] Minatom appears to have much less influence over Khabarovsk Kray than it has over Kamchatka Oblast and Primorskiy Kray, probably because the territory has few nuclear submarines awaiting Minatom-funded dismantlement. However, the Radon plant has expanded its capacity in order to accept new radioactive waste from Primorskiy Kray, without any visible protest by Khabarovsk Kray leaders. Khabarovsk State Duma deputies also voted for changes in Russian laws that will allow the import of spent nuclear fuel from abroad, despite Governor Ishayev's

---

[92] "Atomnyye podvodnyye lodki TOF budut remontirovat na Amurskom sudostroitelnom zavode," Vostok-Media, October 9, 2002.

[93] Boris Savelyev, "Korabely Komsomolska-na-Amure vpervyye nachali remont atomnykh podlodok," ITAR-TASS, November 9, 2003.

[94] Belova, "Vostrebovany navechno...."

[95] Ibid.

[96] Svetlana Podznoyeva, "Na nashem 'Radone' ne prolyut ni odnogo rentgena," *Tikhookeanskaya zvezda*, March 15, 2002.

[97] NTV, October 3, 2001; in "Forest Fires Threaten Nuclear Waste Burial Site in Khabarovsk," FBIS Document CEP20011003000294.

[98] "A Ishayev—protiv!" *Tikhookeanskaya zvezda*, June 13–20, 2001, p. 2.

position against it. Ishayev appears to emphasize local economic gains over other considerations, apparently encouraging Amurskiy Zavod's efforts to lease a nuclear submarine to India.

Amur Oblast

Amur Oblast is home to the Ukrainka heavy bomber base, one of two such bases in the country, and a former central nuclear weapons storage facility at Malaya Sazanka. There are approximately 480 nuclear warheads stored there for use on the 21 Bear H16 and 24 Bear H6 heavy bombers.[99]

Military bases in Amur Oblast have suffered from power cut-offs and a lack of employment opportunities in the territory, although the recent completion of the Bureyskaya hydroelectric station in the oblast should alleviate the former problem.[100] The Ukrainka Airbase too suffers from a severe lack of funds. While it reported that it had more fuel for its bombers in 2002 than in the previous year, it nevertheless did not have enough to train new pilots. In fact, although the majority of the pilots at Ukrainka are young (the average age is 27–28), new flight school graduates "practically do not fly."[101] The airbase also has problems in the social sphere: more than 500 base officers have no apartments. Base officials blame the large number of Chinese in the city for the high rents there, which mean that the base has to pay over 100,000 extra rubles (over $3,000) per retiree in order to move him off the base.[102] The base has sought sponsors for airplane crews, but it the assistance obtained has been insufficient.[103] Amur Oblast also faces an ever worsening drug problem. The long border with China has contributed to the difficulties in fighting the import of psychotropic drugs.[104]

The federal government has not had much success helping Amur Oblast. According to *Kommersant*, federal representatives have said that their most important job is to maintain the political stability of the oblast, because it is home to the Svobodnyy Cosmodrome and Ukrainka Airbase.[105] Oblast politics, however,

---

[99] Weapon numbers determined by bomber carrying capabilities; bomber numbers from START I MoU, July 31, 2001.

[100] "Putin: Bureyskaya GES dolzhna stat fundamentom ekonomiki Dalnego Vostoka," ITAR-TASS, July 9, 2003.

[101] Oleg Pochinyuk, "Boyegotovnost—glavnyy kriteriy," *Krasnaya zvezda*, November 11, 2002.

[102] There have also been difficulties maintaining aviator morale: the base has not even been able to afford newspapers. Pochinyuk, "Boyegotovnost—glavnyy kriteriy."

[103] The desperation of officers at an unrelated Amur Oblast military enterprise is indicative of conditions at Amur military bases: the men went on a hunger strike after having not been paid in 16 months "Provedya golodovku, ofitsery voyennogo predpriyatiya dobilis vyplaty denezhnykh okladov," Polit.ru website, September 7, 2001, <http://www.polit.ru/>.

[104] "V Amurskoy oblasti za 5 mesyatsev 2002 goda iz nezakonnogo oborota izyato 2,5 tonny narkoticheskiykh veshchestv," Vostok-media, June 26, 2002.

[105] Irina Kholmskaya, Svetlana Mayorova, "Amur Oblast," *Kommersant*, March 20, 2001.

have been anything but stable, and the actions of federal officials have not helped the situation. In the 2001 gubernatorial election, for instance, Presidential Representative Pulikovskiy threw Moscow's support behind sitting Governor Anatoliy Belonogov, despite the poor job he had done in running the oblast. However, the governor lost his post, and Pulikovskiy was widely seen to have "lost" the Amur elections.[106] Federal officials are correct, however, in asserting the importance of political stability of Amur Oblast. It is extremely difficult to isolate an installation like Ukrainka from the poverty, drug problems, and criminality in the surrounding territory.

Primorskiy Kray

Primorskiy Kray, the territory that borders both North Korea and China, is home to a number of naval bases, fresh nuclear fuel storage facilities, 30 decommissioned submarines (three of which have damaged reactors), and large amounts of spent fuel and radioactive waste. The nuclear facilities are concentrated on and adjacent to the Shkotovo Peninsula, to the southeast of Vladivostok, about 150 kilometers from North Korea. Primorskiy Kray has been the recipient of large amounts of foreign aid, spent chiefly on submarine dismantlement and construction of the politically controversial *Landysh* liquid radioactive waste processing facility.

The Zvezda Far Eastern Shipyard

Zvezda, located approximately 25 km east of Vladivostok in the city of Bolshoy Kamen, is the primary facility responsible for the repair and dismantlement of Pacific Fleet nuclear submarines and the principal repair facility for exported Russian submarines. The shipyard faced great economic difficulties in the early 1990s, but has become solvent through the highly successful SSBN dismantlement program financed by the U.S. Department of Defense.[107] This foreign assistance program exemplifies successful cooperation with regional actors and the foreign donor's ability to flexibly rework its program in response to local conditions. Primorskiy Kray politicians lobbied together with the shipyard director to push the U.S. program through federal offices in Moscow. In 1995, Zvezda received $6 million worth of equipment from the United States, yet scrapping was slow, since

[106] This was despite the fact that Pulikovskiy characterized both candidates as "tarnished" (obgazhennyye). Igor Verba, "Vertikal vlasti, ili shest dlya golykh ambitsiy," *Nezavisimaya gazeta*, March 23, 2001; Oleg Zhunusov, "Polnomochnyy ili politicheskiy?" *Izvestiya*, May 12, 2001.

[107] In order to facilitate dismantlement, the CTR program has also provided defueling facilities and low-level radioactive waste processing and spent naval fuel storage facilities. "CTR Program: SSBN Dismantlement Project," April 2, 2002.

the shipyard could not afford to operate the equipment.[108] In 1996, the United States started to contract Zvezda for dismantling SSBNs directly: the CTR program has paid preset amounts after each stage of the dismantlement process is completed.[109] The shipyard has prospered under this system, and dismantlement work accelerated.[110]

The main employer in Bolshoy Kamen, Zvezda Shipyard, has often been the focus of local political and environmental disputes. After Zvezda was designated the START-I declared dismantlement facility for the Pacific Fleet, Moscow decided to handle radioactive waste at the facility as well. While the shipyard, and thus the city, depends on dismantlement money for survival, and dismantlement implies dealing with spent nuclear fuel and liquid and solid radioactive wastes, this decision became and remains a hot political issue. Territorial and federal officials fought over the Japanese-funded *Landysh* liquid radioactive waste treatment facility project, while the foreign donors failed to pursue the local organizations that might have lobbied on behalf of the venture, leading to costly delays. Although the facility is complete, local politicians periodically threaten to obstruct operations. While efforts to improve public relations regarding the facility have shown some fruit of late, all concerned would have benefited if more attention had been paid to gaining local support years earlier.[111]

While the United States is funding SSBN dismantlement, Japan is the only country helping fund SSN or SSGN dismantlement in the Russian Far East to date: the first contract for dismantlement of a submarine with Japanese money was signed in November 2003.[112] According to Russian plans, Zvezda is supposed to dismantle all Pacific Fleet SSNs and SSGNs (only 13 of 55 of which have been defueled), which constitute a proliferation threat due to the large amounts of highly enriched uranium contained in their spent fuel. Territorial politicians are very

---

[108] James Clay Moltz, "Conditions At Bolshoi Kamen and Problems of CTR Implementation," Trip Report to Center for Nonproliferation Studies, Monterey Institute, March 1996.

[109] U.S. General Accounting Office, Nuclear Nonproliferation: Security of Russia's Nuclear Material Improving; Further Enhancements Needed, GAO-01-312 (Washington, D.C.: February 2001), GAO website, <http://www.gao.gov/new.items/d01312.pdf>.

[110] By 2007, CTR plans call for the dismantlement of 18 SSBNs at Zvezda. "CTR Briefing, CTR Program: SSBN Dismantlement Project," December 2000, as cited in Jon Brook Wolfsthal, Cristina-Astrid Chuen and Emily Ewell Daughtry, *Nuclear Status Report: Nuclear Weapons, Fissile Materials, and Export Controls in the Former Soviet Union* (Monterey, CA and Washington, D.C.: The Monterey Institute of International Studies and the Carnegie Endowment for International Peace, June 2001), p. 50.

[111] For a detailed history of the assistance project, see Cristina Chuen and Tamara Troyakova, "The Complex Politics of Foreign Assistance: Building the *Landysh* in the Russian Far East," *The Nonproliferation Review* 8 (Summer 2001), pp. 134–149.

[112] Interview of Japanese Foreign Ministry officials cited by Dr. Nobumasa Akiyama, Hiroshima Peace Institute, Japan, in correspondence with author.

concerned by the environmental threat posed by these vessels, some of which are nearly 40 years old.[113]

The availability of costly metals and lucrative contracts at Zvezda has made it a magnet for criminals. According to some sources, former Zvezda Director Valeriy Maslakov's stepson was involved with some criminal groups profiting from the sale of plant scrap metal.[114] Maslakov was trying to put a stop to all scrap metals sales from the yard, but on September 17, 2001, he was murdered by his stepson in a drunken rage.[115] Others have suggested that Maslakov's stepson was in fact nothing but a stooge, used to cover up the tracks of the real killers.[116] In an earlier, even more disturbing incident, a Zvezda deputy director was kidnapped in 1993 by a mafia group trying to seize control over the shipyard.[117] The group attempted to force the deputy, Aleksandr Makarov, to set up an internal network loyal to the mafia. He refused to cooperate and was killed. The criminals were not caught until 1999.[118] Zvezda and its workers are likely to come under increasing stress as SSBN dismantlement nears completion, around the year 2007.[119]

Primorskiy Kray Naval Bases

The climate of violence and the desperation faced by those serving in the regional branches of the armed forces, including those handling nuclear weapons and reactors, are illustrated by suicide epidemics. In June 2000, three Navy men killed themselves within a week. In November 2001, another pair died. One seaman on the destroyer *Admiral Panteleyev* hung himself in mid-November; five days later an officer on the *Irkutsk*, an Oscar II-class SSGN at Zvezda for repairs, shot and killed himself.[120]

As elsewhere in Russia, drug abuse is a significant problem on Primorskiy Kray naval bases, even in the elite nuclear submarine force. It is even more likely

---

[113] For instance, there are four decommissioned November-class SSNs in the Pacific Fleet, which were constructed between 1955 and 1963. Discussions with Pacific Northwest National Laboratory scientist, May 2000.

[114] CNS interview with DTRA official, October 2001.

[115] Elena Osokina, "Ubit direktor oboronnogo zavoda 'Zvezda' v primorskom gorode Bolshoy Kamen; po podozreniyu v prestuplenii zaderzhan yego syn," ITAR-TASS, September 17, 2001.

[116] Interviews of Bolshoy Kamen residents in 2003 by Russian researcher, related in interview with author, January 29, 2004.

[117] Cited in James Clay Moltz, "Russian Nuclear Regionalism: Emerging Local Influences over Far Eastern Facilities," *NBR Analysis* 11 (December 2000); pp. 45–46; original report in Vadim Bertsov, "Bandity predlagali 'kryshu' oboronnomu zavodu," *Kommersant*, March 19, 1999.

[118] Bertsov, "Bandity predlagali 'kryshu' oboronnomu zavodu."

[119] The CTR program calls for the final two Delta submarines to be dismantled in 2007. CTR Briefing.

[120] Andrey Kalachinskiy, "Zhizn ofitsera razbilas o lodku," *Novyye izvestiya*, November 28, 2001.

among men assigned to decommissioned submarines, since these vessels do not have the sponsors subsidizing their livelihood.

Economic desperation has also led naval personnel to resort to crime. The thefts not only harm the safety and readiness of Russia's submarine fleet, but also open the door to possible thefts of nuclear materials and endanger the environment. For instance, in June 2001, a lieutenant and a seaman from the Vladivostok garrison attempted to steal 70 Pacific Fleet submarine air filters, which they planned to sell in China.[121] The threat of insider theft remains acute so long as naval personnel and guards continue to make very low salaries. The five guards asphyxiated in a March 2000 attempt to steal metal from a partially dismantled Yankee-class SSBN had been earning just $25 a month.[122] Shipyards also face the dangers of insider thefts. A Zvezda engineer, caught in late 1999 with some 3.5 kg of uranium alloy, had evidently stolen the metal sphere and kept it in her garage for quite some time while attempting to find a purchaser. The source of this unenriched uranium metal was never definitively determined.[123]

The problem of insider thefts may increase if a lucrative market for militarily sensitive equipment develops. The North Korean withdrawal from the Non-Proliferation Treaty in January 2003 is especially worrisome in this respect. North Koreans have been caught in the past attempting to obtain submarine training schedules.[124] The local Russian press has printed claims that Chinese have tried to recruit Russian defense industry workers to steal particular items in their factories.[125]

One way to prevent theft is to step up security, aimed at both insider and outsider threats. The U.S. DOE, under the Material Protection, Control and Accounting (MPC&A) Program, has provided security upgrades to several sites in Primorskiy Kray.[126] It is important that attention to the personnel manning these

---

[121] Lyubov Troyanovskaya, "Flotskiy Klondayk," *Yezhednevnyye novosti*, June 19, 2001.

[122] "Gibelnaya Chazhma," *Vladivostok* March 22, 2000; Valeriy Aleksin, "Nechastnyy sluchay," *Nezavisimaya gazeta* March 23, 2000; Denis Demkin, "Moryaki zadokhnulis v spisannoy podlodke," *Kommersant*, March 22, 2000; "Pogibli po prichine slaboy distsipliny," *Vladivostok*, April 5, 2000.

[123] "Na dalnovostochnom zavode po remontu atomnykh podvodnykh lodok 'Zvezda' ne vyyavleno faktov khishcheniy urana," Interfax, September 3, 1999; Georgiy Kulakov, "Prodavtsy urana prinyali militsionerov za banditov," *Kommersant*, September 2, 1999.

[124] *Segodnya*, October 21, 1994, in Alexander Zhebin, "A Political History of Soviet-North Korean Nuclear Cooperation," in James Clay Moltz and Alexandre Mansourov, eds., *The North Korean Nuclear Program: Security, Strategy, and New Perspectives from Russia* (New York: Routledge, 2000), p. 36.

[125] Boris Savelyev, ITAR-TASS, February 5, 1999.

[126] The programs include work on the Pacific Fleet Navy Refueling Ship PM-74 and two nuclear fuel storage facilities on the Shkotovo Peninsula. An expanded MPC&A agreement was signed in September 2000. U.S. Department of Energy MPC&A Task Force Personnel Presentation, Monterey, CA, August 6, 1999; Rear Admiral Nikolay Yurasov et al., "Upgrades to the Russian Navy's Fuel Transfer Ships and Consolidated Storage Locations," Partnership for Nuclear Security: United States/Former Soviet Union Program of Cooperation on Nuclear Material Protection, Control, and Accounting, September 1998; "V

facilities be maintained, so that they continue to receive the necessary training and are paid fully and on time. Otherwise, upgrades to the physical plant will do little to protect sensitive materials.

Primorskiy Politics

The territory has been politically unstable for much of the past decade, with undeniably negative effects on local nuclear facilities. Local leaders have vying patrons in Moscow and have played out their disputes on a national stage. Yevgeniy Nazdratenko, governor of the territory for eight years, resigned in early 2001, under strong pressure from Moscow. The current governor, Sergey Darkin, was elected in a closely fought race against the deputy to Far Eastern Presidential Representative Pulikovskiy in the summer of 2001. The territory's political woes have continued: Primorskiy Kray was unable to elect a new legislature in December 2001, because insufficient numbers of voters went to the polls,[127] and there remains still a political standoff between the governor and the mayor of Vladivostok, a Nazdratenko supporter.[128]

Despite its political difficulties, there are a few examples of positive collaboration between Primorskiy Kray politicians and local nuclear facilities. For instance, local city government and territorial government officials lobbied Moscow together with the former director of the Zvezda Shipyard, the START-designated SSBN dismantlement facility for the Pacific Fleet, in support of the CTR submarine dismantlement program.

However, most of the impact Primorskiy Kray has had on its host facilities has been negative. In addition to political problems and the inability of territorial and federal politicians to cooperate—even in the nuclear sphere, where there is much at stake—the territory suffers from power shortages, economic difficulties, and a high

---

Primorskom kraye otkryty dva khranilishcha radioaktivnykh otkhodov," *Nezavisimaya gazeta*, September 2, 2000, p. 2; "Secretary Richardson hails completed security upgrades at ceremony in Russian Far East," September 1, 2000, United States Department of Energy website, <http://www.energy.gov>.

[127] Local experts attribute this to the governor's administration, which ignored the race, and to Presidential Representative Pulikovskiy's first deputy, Aleksandr Drozdov, who declared that one fifth of the candidates running for election had criminal ties. The legislature was finally elected in June 2002. Sergey Akulich, "Pervyye porazheniya Sergeya Darkina," *Vladivostok*, December 11, 2001; Andrey Kalachinskiy, "Skolko 'banditov' budet v Dume Primorya?" *Novyye izvestiya*, November 27, 2001; Igor Nikitin, "ZakS budet!" *Zolotoy rog*, June 6, 2002.

[128] Some local journalists have claimed that Nazdratenko supporters and supporters of erstwhile Nazdratenko opponent State Duma deputy Viktor Cherepkov have allied against Darkin. The Primorskiy Kray Legislative Assembly, formed in June 2002, has a majority of Nazdratenko and Cherepkov supporters. Cherepkov has been quoted as saying that Darkin's political position is extremely vulnerable. Aleksey Chernyshev, "On chto, protiv Putina?" *Kommersant,* December 19, 2002; Aleksey Chernyshev, "Gubernator Primorya ne boitsya, chto ego vykovyryayut lomom iz kresla," *Kommersant*, June 24, 2002.

crime rate. All of these factors put facilities in the territory at high risk. As noted above, criminal groups, or "mafias," have specifically targeted Zvezda Shipyard, attempting to gain control over some of the lucrative contracts at the facility through threats and contract killings. The danger remains that criminals will find a way to infiltrate the site in future.

Primorskiy Kray's rocky relations with Moscow periodically have made local problems yet more difficult. Although he had some strong supporters in Moscow behind the scenes, Nazdratenko (governor from 1993 to 2001) was often in open conflict with central authorities. He used power outages as a way of holding the territory hostage, thereby forcing Moscow to pay promised subsidies to the territory. He even ordered that power to military facilities and early warning radars be cut off on occasion. In order to lessen dependence on the power concern, Nazdratenko supported the idea of using nuclear submarines to provide electricity, as well as construction of a nuclear power plant; neither of these plans, however, was enacted. After Nazdratenko's ouster, Presidential Representative Pulikovskiy made an unsuccessful attempt to reign in the territory. As in Amur Oblast, his failure to get the candidate he supported elected to the governor's chair weakened, not strengthened, Moscow's authority.[129]

The Ministry of Atomic Energy and Primorskiy Kray

Due to the significant environmental threat posed by the many decommissioned nuclear-powered vessels in the territory, most local politicians appear to set a premium on good relations with Minatom. They tend to support Minatom's projects for the territory, as well as its policies in Moscow. Although Governor Darkin stated that he would hold a referendum[130] before making any decisions regarding construction of a nuclear power plant in Primorskiy Kray, in April 2003 Minatom announced that it planned to resume construction of the Primorsk NPP.[131] Local politicians have also supported Minatom plans at the federal level. For instance, Bolshoy Kamen Mayor Anatoliy Karasev has spoken out in favor of importing spent nuclear fuel, an important Minatom project, via Primorye.[132] It remains to be seen whether or not Minatom will spend more money, or dismantle more boats, in those territories that have given it political support.

The Far Eastern Federal Enterprise for Handling Radioactive Wastes, or DalRAO, subordinate to Minatom, was set up in February 2000 to manage

---

[129] Nor has Pulikovskiy played a visible role in solving local nuclear problems. In contrast, Viktor Cherkesov, the presidential representative to Northwest Russia, sponsors regular meetings with Northern Fleet officials and has a special division in his administration that deals with nuclear issues.

[130] Aleksandr Kartashov, "Gubernator privetstvuyet AES i kritikuyet ZakS," *Vladivostok*, December 26, 2001.

[131] ITAR-TASS, April 23, 2003.

[132] Bolshoy Kamen and the Vladivostok Commercial Port, as well as Pevek and Provideniya, Chukotka, are the four far eastern ports on the official list of ports that may transfer spent nuclear fuel. Vostok-media, December 25, 2003.

radioactive waste sites, coordinate the dismantlement of ships and nuclear submarines, and rehabilitate the environment at naval bases in Kamchatka and Sakhalin Oblasts and Primorskiy Kray.[133] However, many decisions regarding naval dismantlement issues continue to be made in Moscow. Therefore, Primorskiy Kray politicians visit Moscow to voice their concerns regarding submarine dismantlement.[134]

Conclusion (and Issues for Foreign Assistance Providers)

There are several urgent nuclear problems in the Far East region that still require greater attention, both from Moscow and abroad. The oldest decommissioned vessels in Russia's nuclear navy pose the greatest danger to the environment, and also have spent nuclear fuel that has become less radioactive with age, thus making theft more possible. However, most of these vessels are general-purpose nuclear submarines, and thus do not fall under the U.S. CTR assistance program. While Minatom is requiring shipyards to scrap these submarines, it is not always able to pay for this work. Should Zvezda Shipyard fail to pay its workers, an increasing possibility as the end of SSBN dismantlement nears, the danger of insider thefts of nuclear materials will grow considerably.

There are new regional players likely to become involved in submarine dismantlement and new nuclear construction projects in the next few years. Since 2001, the governments of Primorskiy Kray and Kamchatka Oblast have been required to cooperate with Minatom in developing measures to ensure the safety of dismantlement activities, while the Kamchatka Oblast administration is also supposed to be involved in designing a plan for a solid radioactive waste burial site in the oblast.[135] (In the past, regional administrations were not directly involved in managing nuclear sites.) New entities may become involved in the dismantlement of Pacific Fleet nuclear attack submarines in the near future. On June 27, 2002, the Group of Eight (G-8) states initialed the Global Partnership Against the Spread of Weapons and Materials of Mass Destruction, whereby the G-8 would raise up to $20 billion over the next 10 years to fund nonproliferation projects, mainly in the former Soviet Union. Putin has identified the disposal of non-strategic nuclear

---

[133] A separate organization, Nuklid, coordinates foreign assistance to regional nuclear facilities. Government Directive No. 220-r, "Rasporyazheniye pravitelstva Rossiyskoy Federatsii," February 9, 2000, in *Sobraniye zakonodatelstva Rossiyskoy Federatsii*, No. 7, February 14, 2000, p. 1750; "'DalRAO' zaymetsya razdelkoy korabley i podlodok," *Zolotoy rog*, August 8, 2000.

[134] For example, in December 2001 Governor Darkin met with Minister of Atomic Energy Aleksandr Rumyantsev to discuss liquid and solid radioactive waste treatment issues. Kartashov, "Gubernator privetstvuyet AES."

[135] Viktor Kolesnikov, "Lodki otstoya rzhaveyut i tonut, 'mogilniki'—ne okhranyayutsya, mestnyye vlasti zhdut ot Moskvy ocherednoy federalnoy programmy" (interview of Valeriy Kochetov), *Novaya kamchatskaya pravda* online edition, October 18, 2001, <http://www.iks.ru/~nkp>.

submarines as one of the most important problems facing Russia.[136] So far, however, the only country helping to dismantle non-strategic submarines in the Russian Far East is Japan. Although Japan has indicated that it is concerned about all Pacific Fleet nuclear-powered attack submarines, its negative experience with the *Landysh* project has led it to go slowly in designing its assistance programs. It is therefore unlikely that Japanese assistance will be able to deal with the Pacific Fleet decommissioned submarine problem for many years to come. Other G-8 nations have yet to promise new assistance to the region under the Global Partnership. No nation has yet committed to assisting Russia in the construction of on-shore reactor storage facilities, leaving the long-term location of reactors from dismantled submarines in question. While the U.S. DOE has made some security upgrades at nuclear sites in the Russian Far East, material at these sites remains vulnerable. The joint Norwegian/U.S./Russian/U.K. Arctic Military Environmental Cooperation, or AMEC, program, which has contributed a great deal to the security of Northern Fleet facilities, is considering expanding to the Russian Far East, but no decision on the issue has yet been made due to budgetary concerns.[137]

Although most foreign assistance projects involve agreements between Moscow and foreign heads of state, local actors have a great deal of influence over project success. This influence begins even before the decision to launch a project has been made. A lack of timely information can lead to a good deal of local confusion, and elicit a negative local response. For instance, in August 2000, local journalists chronicled the visit of then-Secretary of Energy Bill Richardson to the Kamchatka Shipyard and assumed that the United States was already undertaking projects at the yard. Local politicians and shipyard workers alike noted that expensive preparations for submarine dismantlement had already begun earlier that year and mistakenly believed that the money was coming from the United States.[138] Richardson's visit and the low profile his team kept during the trip—intentionally or not—only further fueled speculations regarding U.S. intentions. In the end, however, the United States failed to invest in a large dismantlement program proposed by Moscow. The earlier funds had apparently come from Russian sources, perhaps in hopes of attracting U.S. funding. The failure of both Moscow and Washington to explain their plans adequately to the local population, unfortunately, left a wake of regional puzzlement and frustration, and have left locals more wary of any future assistance project.

To date, the greatest deterrent to stealing radioactive materials is the lack of buyers. This situation, however, could change. The only way to combat the problem is to dismantle the submarines and store the nuclear materials securely,

---

[136] "Vystupleniye Prezidenta Rossiyskoy Federatsii V.V. Putina na press-konferentsii po okonchanii sammita 'Bolshoy vosmerki' 27 iyunya 2002 goda, Kananaskis, Kanada," Ministry of Foreign Affairs of the Russian Federation, June 28, 2002, <http://www.mid.ru/>.

[137] AMEC e-mail communication with author, November 8, 2003.

[138] Author's interviews, Petropavlovsk, June 14–16, 2000; "V Vilyuchinske stroitsya tretiy mogilnik dlya gryaznykh atomov," *Vesti*, September 28, 2000, p. 1.

with upgraded physical protection and guard forces that are paid sufficient wages and receive them on time. Failing to send SSNs to dismantlement facilities also increases the chance of an accident that could pollute Far Eastern waters, causing political problems for SSBN dismantlement and other cooperation. While the Russian environment may not be the top priority of those primarily concerned about nuclear proliferation, it is in the interest of any assistance provider to make certain that assistance projects are not seen as contributing to local difficulties in reducing environmental risks. Further, the dismantlement of newer attack submarines eliminates the possibility that they might be sold to other countries, and with them nuclear fuel.[139] Russia's apparent willingness to lease two nuclear-powered submarines to India makes this possibility all the more likely, if the two countries can come to an agreement over terms.

Whether weak or strong, local leaders have a variety of ways to influence local facilities, from signing (or failing to sign) legislation permitting the siting of a floating nuclear reactor in their harbors, to cutting off the electricity at military installations, to promoting exports of nuclear assets. With an understanding of local dynamics, foreign aid donors and Russians alike can develop incentives for regional politicians to assist projects and serve as project advocates, both at home and in Moscow. Improving the security of nuclear weapons and materials located in the Russian Far East requires an understanding of these local dynamics. Russian federal officials and foreigners alike must take into account how regional factors may affect projects in Russia's regions. By studying the territories, many problems can be avoided and new opportunities found.

---

[139] On this topic, see James Clay Moltz and Tamara C. Robinson, "Dismantling Russia's Nuclear Subs: New Challenges to Non-Proliferation," *Arms Control Today* 29 (June 1999), <http://www.armscontrol.org>.

Chapter 7

Nuclear Issues in the Volga Federal Okrug

Ivan Safranchuk

The Volga Federal Okrug consists of 15 federal constituencies.[1] Some of these constituencies, particularly Tatarstan and Bashkortostan, had pronounced separatist tendencies in the 1990s, particularly in the economic sphere. They demanded special rights and privileges from the federal center. Some went so far as to intimate secession, while others ceased remitting taxes to the federal budget altogether. However, these republics and oblasts do not have nuclear facilities in their territory and nuclear issues were not high on their agenda in their relations with Moscow.

Other constituencies of the Volga Okrug, which do house nuclear facilities, have not been engaged in confrontation with Moscow nor have they played the nuclear card in the center-regional politics. To the contrary, federal control and responsibilities in the nuclear area were never questioned. At the same time, close cooperation among nuclear facilities, regional authorities, and the federal government has been developed since 1991 in these oblasts.

This chapter focuses on Nizhniy Novgorod Oblast, which not only has the largest number of nuclear facilities in its territory among the Volga Okrug constituencies (see Table 7.1, p. 136), but also is one of Russia's regions where center-periphery relations were relatively smooth during the rough years immediately after the break-up of the Soviet Union. In addition, it is one of the few regions with a significant nuclear infrastructure located in Central Russia and thus, one could argue, has adjusted to economic and political reforms more rapidly than more distant regions.

This chapter reviews the status and trends of nuclear decentralization through detailed cases studies of Nizhniy Novgorod Oblast, the closed city of Sarov, and one of Russia's largest federal nuclear centers—the All-Russian Scientific Research Institute for Experimental Physics (VNIIEF), which is located in Sarov. The material for this chapter was gathered from personal interviews with VNIIEF

[1] The Volga Federal Okrug includes the following 15 federation constituencies: six republics (Bashkortostan, Mariy-El, Mordovia, Tatarstan, Udmurtia, Chuvashia), eight oblasts (Kirov, Nizhniy Novgorod, Orenburg, Penza, Perm, Samara, Saratov, and Ulyanovsk), and one autonomous okrug (Komi-Permyatskiy).

representatives and local Sarov administrators, the open media, and from the author's personal observations from visiting this closed city.

Table 7.1 Key Nuclear Facilities in Nizhniy Novgorod Oblast

Location	Name	Comments
Sarov	RFNC VNIIEF	One of Russia's two principal nuclear weapons design laboratories
Sarov	Avangard Electromechanical Plant	A serial production plant for assembly/disassembly of nuclear warheads
Surovatikha	ICBM dismantlement plant	
Nizhniy Novgorod	OKBM	Nuclear reactor design bureau (including naval reactor design)
	Krasnoye Sormovo	Produced 2nd and 3rd generation nuclear submarines
	Lazurit Central Design Bureau	Designed nuclear and diesel submarines
	Radon Special Combine	Radioactive waste disposal and storage facility
	NIIIS (Scientific-Research Institute of Measuring Systems)	Produces electronics and control equipment for the nuclear industry
	Atomenergoproyekt	Scientific and research institute involved in nuclear power plant design

Nizhniy Novgorod Oblast: An Economic Overview

Nizhniy Novgorod Oblast covers 76,900 square km and has a population of 3,598,300. As of 2002, the gross regional product constituted 2.2 percent of Russia's GDP. Oblast retail sales, agricultural production, capital investment, and manufacturing comprised 2.2 percent, 1.8 percent, 1.6 percent, and 2.1 percent, respectively, of national totals. Industry accounts for 31.9 percent of oblast output.[2] The primary processing sector includes mechanical engineering (production of automobiles, ships, machine tools, and instruments), power engineering, and elements of the food processing, chemical, and petrochemical industries.[3]

[2] Russia's Regions Investment Statistics, Ministry of Economic Development and Trade of the RF, <http://www.economy.gov.gu>.

[3] *Regiony Rosii 1999*, (Moscow: 1999), p 167.

The largest enterprises in Nizhniy Novgorod Oblast are GAZ, PAZ, and Krasnoye Sormovo. A significant recent characteristic of the region has been the active influx of national financial-industrial groups (FIGs).[4] This, in turn, has fostered a significant redistribution of property. These high profile ownership changes have been accompanied by scandals. For example, when shares for Krasnoye Sormovo were bought up and attempts were made to change the board of directors, the new owner faced opposition from representatives of the Ministry of State Property.[5] In 2000, a similarly nefarious transaction occurred when the private company, Kaskol, acquired another large company in Nizhniy Novgorod, the Sokol aircraft construction plant.[6]

A significant share of the industrial potential of the oblast is oriented toward work for the military-industrial complex (MIC). Military production accounts for one forth of all industrial output in the oblast. As a rule, these large specialized enterprises have trouble integrating into the region's new market economy. This can be attributed less to low levels of business activity than to the fact that a significant number of the enterprises in Nizhniy Novgorod Oblast have been designated as part of the national defense mobilization capacity.[7] Consequently, the oblast administration has been put in the unenviable position of having to wait for approval at the national level regarding the transfer and conversion of idle factories under its jurisdiction.

Arzamas-16: The Soviet Experience

Arzamas-16 was the first of the Soviet closed nuclear cities. It was established in 1946 to house nuclear warhead research, design, and production facilities, which later become known as the All-Russian Scientific Research Institute for Experimental Physics and the Avangard Electromechanical Plant. Throughout the Soviet period, Arzamas-16[8] (population: 84,000) was not only closed for Russian visitors (and, of course, foreign), but its very existence was kept secret. The primary channel of communication was directly with Moscow, rather than with surrounding cities. Residents of nearby settlements suspected that defense activities took place behind the barbed wire, but such operations were never

---

[4] In 2000, there were management changes at the automotive giant GAZ, that was acquired by Sibal, and Krasnoye Sormovo, that was acquired by the holding company, Uralmash-Izhora.

[5] The Ministry of State Property used the state shares to compel decisions that violated court rulings. Galina Kulikova, "'Krasnoye Sormovo': novyy rasklad," *Gorod i gorozhane*, No. 3 (828), January 16–22, 2001, p. 6.

[6] *Nizhegorodskiy Biznes* on-line, June 9, 2000, <http://www.innov.ru/news/nn/062000/09-11.htm>.

[7] *Nizhegorodskiy Biznes* on-line, July 20, 2000, <http://www.innov.ru/news/nn/072000/20-7.htm>.

[8] Another name for the city is Kremlev. It was used for a short time during the transition from an "Arzamas number" to Sarov.

officially or publicly acknowledged. In effect, Arzamas-16 constituted an "island" in the surrounding oblast that was virtually isolated and inaccessible. The situation within the city was typical for Soviet provincial industrial cities with "city-defining" enterprises. The director of such an enterprise was, by virtue of his position, "tsar and god," wielding considerably more influence than local Communist party officials.

Until the end of the 1980s, Arzamas-16 had still not experienced funding shortfalls. The standard of living in the city was even higher than in Moscow. This was reflected primarily in the relatively high average wages. Moreover, the city had a generous program of social benefits (vacations, housing, etc.), which, unlike similar services offered in the capital, were not plagued by corruption. Consequently, the material well-being of the employees of VNIIEF and residents of the closed city was considerably higher than the national average.

But, in the final years of the Soviet Union during the 1980s, when defense orders sharply decreased, the socio-economic situation in Arzamas-16 began to deteriorate. In 1991, a government commission issued a report on the "impoverished condition of the country's nuclear centers" and recommended that the government allocate additional funding to the closed nuclear cities. These recommendations were supported by the Central Committee of the Communist Party and the appropriate directive to the government of the USSR was prepared. Ultimately, however, this funding was not provided.

Sarov: 1991–96

In early 1992, shortly after the Soviet collapse, Russian President Boris Yeltsin visited Sarov. Shortly before the visit, however, prices had been liberalized throughout the country, which dramatically reduced the material well-being of Russian citizens. For the residents of Sarov, these changes were particularly alarming. The unprecedented hard times led local residents to question the government's commitment to VNIIEF and the closed city as a whole. Generous central financing was now an impossibility, which was particularly jarring for residents of Sarov who had enjoyed an exceptionally high standard of living during the Soviet period.

Yeltsin spent only one day in the city, but announced firmly that the Russian government was fully committed to preserving the nuclear center. During the visit, as some of those present remember, he "practically leaned on a warhead" while signing the edict supporting closed cities.[9] This edict included orders to prepare a law "On Closed Administrative-Territorial Formations (ZATOs)." This law was subsequently prepared and adopted in 1992. In accordance with Article V of the law, all taxes (local and federal) collected within closed cities were slated to

[9] Conversation with heads of OKP-276 (professional union) of VNIIEF, Sarov, January 17, 2001.

remain in the local budget. After the law was amended in 1996, the City Duma gained the right to offer tax privileges.

Notwithstanding these political and normative gestures, the standard of living in the city did not significantly change from 1991 to 1996. Although VNIIEF transferred the bulk of its social assets and services to the city administration, this was mostly a formality, since the city budget was formed almost exclusively from tax receipts from VNIIEF and Avangard.[10] In practice, social spending remained covered by the institute. Therefore, the formal transfer of social services barely altered relations between local authorities and the institute. With respect to day-to-day affairs, VNIIEF employees were living much worse than they had during the Soviet period. The system of government subsidies virtually unraveled. As a result, the city infrastructure deteriorated. Nevertheless, remaining reserves from the "golden Soviet era" still insulated the city from many of the problems experienced across the country and encouraged the residents to clutch to false hopes of a turnaround.[11]

The central government attempted to redress the problems confronting VNIIEF and the city by showering both with high-level attention. Then-Minister of Atomic Energy Viktor Mikhailov, who came up through the ranks as a nuclear weapons scientist, sought to reassure the residents that the federal government would continue to look after their well-being.[12] The people of Sarov considered him one of their own, and he was a relatively effective lobbyist for the institute and the city. Moreover, for three successive years (1995–97), Prime Minister Viktor Chernomyrdin visited Sarov.

Still, the level of crime in Sarov remained low. But there was also very little business activity. The director of VNIIEF remained the most prominent figure, not only at the institute, but within the city as a whole. Even while material conditions declined precipitously, there was a sense among the local population that the barbed wire fence encircling the closed city was protecting the citizens from a threatening new reality outside.

During the initial period following the Soviet collapse, the main problem confronted by Sarov was the declining level of government subsidies. Throughout the 1990s, the problem of indebtedness of the federal budget to VNIIEF persisted. As a rule, Minatom financed the institute almost completely, but with delays. The head of the institute traveled to Moscow to work with Minister Mikhailov to lobby the government for relief. However, wages went unpaid for months. Due to the unreliable financing on the part of the Ministry of Finance, the transfer of funds to the VNIIEF account for a given year was typically delayed until the end of the first quarter of the following year. As a result, wage payments were delayed by 2.5 to 3

[10] Conversation with deputy of the Sarov City Duma, January 17, 2001.

[11] Conversation with heads of OKP-276 (professional union) of VNIIEF, Sarov, January 17, 2001.

[12] Nuclear weapons scientists also call themselves "bombodely"(bomb-makers).

months.[13] Throughout this difficult period, the Communist Party remained the dominant political force in Sarov. Memories of "the good old days" and the "graying" of the VNIIEF provided the Communist Party with a solid footing in the city. However, the city was subordinate to the center, and maintained a clear orientation toward the pro-government party. Thus, in the 1993 and 1995 parliamentary elections, support for the Communist Party was shared with local endorsement of the pro-government parties, "Russia's Choice" and "Our Home Russia," respectively. Similarly, Boris Yeltsin won the majority of the vote in the 1996 presidential election in the city. This trend remained strong in subsequent years. In 1999, during parliamentary elections, a significant portion of the votes was collected by the pro-Putin "Unity Party," and in the March 2000 presidential elections, Vladimir Putin himself garnered most of the local vote. Nonetheless, throughout all the 1990s, only the Communist Party retained a stable, although not decisive, position in Sarov.[14]

Between 1991 and 1996, Sarov continued to exist as a city subordinate to the center. Economic ties with the oblast expanded (for the purchase of foodstuffs and the like), but the city existed outside of the oblast's economic space. It remained an island that was much closer to Moscow than to Nizhniy Novgorod Oblast.[15] This situation did not completely suit regional authorities. Oblast Governor Boris Nemtsov planned to take an active role in shaping the business of Sarov and VNIIEF. The governor's office criticized the scant reforms implemented by VNIIEF, dismissing the significance of the conversion programs. Yet the governor, too, offered neither policy direction nor substantive relief.

During this period, relations between central and regional authorities concerning VNIIEF activities remained unchanged. There were no inklings of separatism in Nizhniy Novgorod Oblast. In fact, it appeared that ties to the center were strong, especially given Governor Nemtsov's close association with President Yeltsin.

Only once, at the very beginning of the 1990s (in 1992-93) did debates unfold concerning Sarov that had the potential to create a crisis situation. In 1992, "Zemlya," a regional movement registered in adjacent Mordovia, proposed placing Sarov under its administrative jurisdiction. Although 97 percent of the city and the entire institute are located on Mordovian territory, both had historically been placed under the jurisdiction of Nizhniy Novgorod Oblast administration.[16] The

---

[13] In 1996, during Yeltsin's re-election campaign, Sarov managed to correct its financial position somewhat. The wage arrears were almost completely paid up.

[14] The deputy of the State Duma from the single-mandate region that includes Sarov is Nikitchuk, a communist; the Sarov deputy to the Nizhniy Novgorod Oblast Legislative Assembly is also a communist.

[15] Conversation with S. V. Subbotin, docent of the Political Science Department of Lobachevskiy Nizhniy Novgorod University, Nizhniy Novgorod, January 19, 2001.

[16] The land officially remains part of Mordovia, that is, the borders include this territory. The 3 percent of the territory that lies in Nizhniy Novgorod Oblast includes only a small part of the city. The entire territory of VNIIEF is officially on Mordovian soil, on the territory of the Mordovian Nature Reserve. However, it falls under the administrative

"Zemlya" movement petitioned for the transfer of the territory of VNIIEF and a large section of the city of Sarov, to the jurisdiction of the Republic of Mordovia. This effort apparently was not motivated by political or nationalistic reasons. Rather, the movement championed its cause on cultural grounds under the banner of "Collecting the Lands" of Mordovia.[17] At the same time, several senior officials of VNIIEF took seriously the possibility that Mordovia, riding a wave of heightened national consciousness, could maneuver to take control of the nuclear center, with unpredictable consequences. In any case, after Moscow intervened, the issue of transferring jurisdiction was put to rest. Mordovian authorities expressed no claims in this regard. Zemlya's slogans (perhaps with "help" from local authorities under pressure from Moscow) did not become widespread, and the issue died of its own accord.

Sarov: 1997–2000

Sarov's relationship with different levels of the federal administration changed significantly in 1997. From 1997 to 2000, new ties were established with central, regional, and local powers. New bonds also were forged between Sarov and the VNIIEF leadership. Until 1997, one could speak only of relations between Sarov and Moscow, as formal ties between Sarov and Nizhniy Novgorod Oblast were practically nonexistent. After 1997, however, there were changes in all directions (Moscow-VNIIEF, Sarov-Nizhniy Novgorod Oblast, Sarov-VNIIEF). The primary impetus was a new economic initiative[18] that resulted from the granting of "off-shore" status to closed cities by the federal government in 1996.

In February 1997, the Sarov Duma passed a law that created a tax-free investment zone in Sarov. The "founding fathers" of the zone were Governor Nemtsov, Sarov Mayor Gennadiy Karatayev, and VNIIEF Deputy Director Sergey

---

control of the Nizhniy Novgorod Oblast. Over several decades, the perimeter of the city and institute have gradually increased (by approximately by 20 percent). This growth has occurred at the expense of Mordovian land, which was subsequently transferred to Nizhniy Novgorod (formerly Gorky) Oblast's administrative control.

[17] At present, Moscow and Sarov have differing attitudes toward this story. It is a commonly held viewpoint that the Zemlya story was nothing serious, since they had limited influence in Mordovia and the republic's authorities did not support their claims for returning the territory, either directly or indirectly. Conversation with high-ranking VNIIEF employee, Sarov, January 18, 2001.

[18] In 1997 and 1998, a number of events took place that had unquestionable influence on the fate of the nuclear centers. For example, in 1997, Radiy Ilkayev was appointed the new director of VNIIEF. In 1998, nuclear weaponeer Mikhailov, who was closely tied to VNIIEF, was succeeded as Minister of Atomic Energy by Adamov, a "power engineer," which all those involved in the nuclear weapons complex took as a bad omen. But the greatest influence on the existence of Sarov and VNIIEF from 1997 to 2000 was the operation of the investment zone, which led to the establishment of new mutual relations concerning Sarov on all levels of power, rather than various personnel shuffles.

Krysov.[19] Boris Nemtsov insisted on pushing through conversion programs at VNIIEF. In December 1996, he had visited Sarov and signed a protocol that created the company, VNIIEF-Konversiya (or "Conversion Fund"), to manage this process.

In addition to VNIIEF-Konversiya, two other funds were established: the fund for social development, and the fund for fostering law and order. For each fund, a board of trustees was formed. The board confirmed the projects and amounts of financing for each of the funds. The board of trustees of the social fund included the entire City Duma, the mayor, and the deputy mayor for economic affairs.[20] The board of trustees for the conversion fund was headed by the director of VNIIEF and included representatives from VNIIEF, Avangard, the City Duma, and the mayor's office. The heads of local law enforcement agencies and several City Duma deputies comprised the board of trustees for the law enforcement fund.

The Sarov investment zone took shape in April 1997 and by the second quarter, the zone had registered 30 enterprises (the majority were from Nizhniy Novgorod). The funding received from the investment zone was disbursed unevenly between the city budget and the three funds.[21]

However, the operation of the investment zone did not render the city budget completely self-sufficient. Throughout the period, the city administration continued to receive federal subsidies that covered about one-third of municipal expenses.[22] Moreover, Sarov consistently was a leader in terms of its receipt of federal funding (even among the other nuclear closed cities). In addition, VNIIEF and Avangard continued to provide the largest tax payments to the city budget, equalling approximately half of the city's overall tax receipts.

---

[19] Sergey Krysov was a protégé of Boris Nemtsov. However, Krysov's push for conversion at VNIIEF met resistance and he had to leave his post of deputy director in May 1998. See details later in the chapter.

[20] In 2000, after the City Duma elections, a scandal erupted around this fund's board of trustees. In accordance with the social development fund's charter (all the funds' charters were accepted by the City Duma), the initial board of trustees was to be formed by the founders—the City Duma. In accordance with the charter, changing the makeup of the board of trustees, holding elections, and changing its management bodies is the exclusive province of the board of trustees. Thus, the initial board of trustees was unaffected by the results of the City Duma elections. New deputies were not automatically included in the board, and those who had been voted out of office retained their power in the board. *Noviy Gorod*, January 18, 2001.

[21] The controversy over the investment zone and Nizhniy Novgorod Oblast's position on the special tax treatment for Sarov is discussed later in the chapter. See section "Sarov and Nizhniy Novgorod Oblast."

[22] Conversation with high-ranking official in Sarov city administration, Sarov, January 18, 2001.

VNIIEF and Moscow

Wage arrears were the primary bone of contention between the center and VNIIEF throughout 1997-2000. In 1997, VNIIEF confronted an especially acute problem of wage arrears. On September 16, 1997, the workers at VNIIEF staged a strike in Sarov to protest the fact that the Ministry of Finance remitted payment for only half of the defense order fulfilled by the institute. This debt to the institute for wages exceeded the three-month payroll fund. The workers' collective at VNIIEF complained that "through its actions, the government of Russia was complicit in the collapse of work to support the country's defense capabilities and the safe use of the country's nuclear weapons."[23] In response, Radio Rossii reported that Prime Minister Viktor Chernomyrdin had agreed to release 150 billion rubles to settle wage arrears at Russia's two nuclear weapons design centers (VNIIEF and VNIITF).

Yet, Moscow's promises did not provide sufficient relief, as the wage arrears to VNIIEF by now were virtually equivalent to the entire 1997 state defense order.[24] Repeated delays in salary payments only exacerbated tensions between the workers and the VNIIEF administration. Tensions also were aggravated because municipal civil servants seemed to be living well by comparison, as they had received steady pay increases. In fall 1997, for example, while institute workers battled to collect back wages, the mayor of Sarov signed a directive that ordered a pay increase for low-level civil servants (ranks one through five) to compensate for inflation.

Compounding problems was the stratification of the VNIIEF workforce. The division cut along the lines of those who were paid little and irregularly, and those who were compensated well and on time. There were few employees of the institute who were able to secure additional commercial employment in the city. These people were mostly those involved in conversion programs. They generally did not depend on the state's timely payment of defense orders in order to support themselves.

All institute employees traditionally received small portions of their monthly salaries in advance from the administration, but, in order to do this, it was necessary to arrange additional loans from commercial banks (before VNIIEF and the city administration developed their own loan relationship with low interest

---

[23] On average, between 1997 and 2000, the government paid VNIIEF for only 70 percent of the defense order. But this was not particularly bad, when compared to other military-industrial complex enterprises, including ones located in Nizhniy Novgorod Oblast. As a rule, until 1998, MIC enterprises received payment for 5-10 percent of the state defense order. The management of VNIIEF did not feel that such comparisons were appropriate, however. In 1998 at a joint press conference between Ivan Sklyarov and VNIIEF Director Radiy Ilkayev, the latter stated, "...when you say we were paid for 70 percent of the state defense order, everyone gets upset and is jealous, since many receive only 5 or 10 percent. In my opinion, the government must have priorities, and in my opinion, nothing is more important in the defense business than work on nuclear weapons." *Gorodskoy Kuryer*, No. 51, 1998.

[24] Wage arrears totaled 54.8 billion, while the state defense order totaled 52 billion rubles.

rates). According to Radiy Ilkayev, the director of VNIIEF, the interest paid on these loans sometimes equaled the institute's total expenditures on materials and equipment. Moreover, Minatom did not compensate the institute for interest payments, thus deferring to VNIIEF to cover shortfalls with income generated by conversion projects.

In 1998, the debt problems intensified. VNIIEF workers undertook various protest actions. Some traveled to Moscow to stage protests directly outside of Minatom, the Ministry of Finance, and the State Duma.[25] Throughout the period, relations among the institute administration, the workers' unions, and ordinary workers were very difficult.

The institute administration could not neglect the demands of the workers' collective. The recurring claims by the institute management that the difficult financial situation had been caused by under-financing from Moscow had the effect of channeling worker dissatisfaction toward the central government. The agitation of the "masses," in fact, complemented the institute's efforts at lobbying the federal government. According to Ilkayev, the strikes and demonstrations made it possible to "extract" additional funds from Moscow, albeit belatedly. Nevertheless, the administration tried to contain economic protests and prevent them from fueling a political confrontation. Institute administrators took care to direct their wrath at federal bureaucrats at the Ministry of Finance and to avoid criticizing the country's political leadership. In fact, the administration occasionally went out of its way to issue statements supporting the president and prime minister in their efforts to broker economic reform.

At the same time, the workers' union, which in general supported the VNIIEF administration, did not always enjoy support from rank-and-file employees, and sometimes found it necessary to challenge the administration on their behalf. In April 1998, the council of the workers' collective issued a number of demands to the administration and threatened to strike. In May, the administration announced that these demands would not be met. As a result, a conflict resolution commission was created in VNIIEF, and relations between management and the workers devolved into a collective labor dispute. This situation continued for more than a year, but never resulted in a strike at the institute.[26]

At the federal level, following the appointment of Prime Minister Sergey Kiriyenko in 1998, a dispute arose over the practice of extending privileges to

---

[25] The picketing of Minatom, according to the chairmen of the institute's unions, was largely symbolic and sometimes political. On one hand their "native" ministry did much to extract "nuclear money" from the Ministry of Finance. On the other hand, protest actions were primarily supported by the left: the Communist Party, the Agrarian Party, and Narodovlastiye. When Adamov became minister, he did not really form a relationship with these groups.

[26] In May 1999, during preparations of collective working agreement at VNIIEF, the council of the workers' collective tried to include in the draft agenda a vote of no confidence in the director of the institute. However, this issue was not included in the final agenda. Moreover, this was not the first time, as in the early and mid-1990s some delegates to the council had regularly tried to arrange a vote of non-confidence in former VNIIEF director Belugin.

closed cities. In May 1998, during a visit to Sarov, Kiriyenko made the following announcement:

> I would particularly like to take this opportunity to review the general situation in closed cities, and especially the tax privileges that exist here. This is a serious issue for the government. That it is necessary to support mechanisms for internal earnings for the development of closed cities goes without saying. That this often happens with an enormous amount of internal misuse is also obvious. According to our estimates, last year the state lost funding on the order of 4 to 5 trillion rubles as a result of the tax privileges extended to the nuclear closed cities. Therefore, it is important not to overstep the bounds of reason, and we need an economic analysis.[27]

Kiriyenko concluded the visit by suggesting that the federal government was preparing to reconsider the benefits of the special tax exemptions provided to Sarov. "First, let's count everything accurately, and then we will calmly decide what we can do better."[28]

In 1998, Minister of Atomic Energy Adamov proposed settling debts to VNIIEF workers by reallocating revenue generated by the investment zone. This proposal precipitated a protracted conflict among the local authorities (mayor and city Duma), the administration of VNIIEF, and Minatom. In short, the new minister protested the formation of the investment zone. He wanted all the money to pass through the center and be disbursed by the ministry within the industry. To the extent that the investment zone could make a contribution, it should be aimed at providing social relief strictly for VNIIEF and Avangard. Accordingly, Adamov demanded that the city administration earmark funds for VNIIEF and Avangard.[29] Yet, the minister did not join forces with opponents of the investment zone. Rather, he supported the representatives of the closed city when this issue was under review at the federal level. In particular, he endorsed the perpetuation of special privileges for Sarov and Snezhinsk in 2000. Moreover, in 2000, during a discussion of the following year's budget, Adamov defended a plan under which the Sarov budget would receive the same level of financing that the city had received from the investment zone, with these resources coming from the federal budget.[30]

---

[27] *Gorodskoy Kuryer*, No. 51, 1998.

[28] This is how the director of VNIIEF R. Ilkaev cited Kirienko at his press conference after the prime minister departed from Sarov. *Gorodskoy Kuryer*, No. 51, 1998.

[29] In March 1998, Adamov arrived in Sarov and tried to conduct these meetings in the traditional style: calling the city "fathers" on the carpet. They came to the minister in one of the VNIIEF buildings, and waited a long time for an audience.

[30] Despite complaints leveled by Adamov, the city administration, having received most of their income from the operation of the investment zone, "shared" revenues with the city-forming enterprises. For example, in 1997, VNIIEF received from the city 22 extensions for paying taxes. For more than one year, Avangard was exempted from paying property taxes (8 billion rubles per year). In addition, Uran, an enterprise of the Sixth Main Directorate of Minatom, was registered in the investment zone.

VNIIEF and the City of Sarov

The authority of municipal leaders and the management of VNIIEF was formally "divorced" in 1992 when most of the institute's social services were transferred to the city jurisdiction. Yet, the VNIIEF director remained the *de facto* "master of the city," with responsibilities that included management of the municipal economy. The intimacy of the institute's involvement in municipal affairs was summed up by the VNIIEF director, who quipped that "one person in VNIIEF hears a report every day of how many traffic accidents there are [in Sarov], how many murders…."[31]

But the city began to gain in economic clout. The creation of the investment zone provided an infusion of capital into city coffers that provided municipal authorities with an independent source of revenue.[32] This money flowed into the city budget regularly and without delay, compared to the constant financial delay in payments to VNIIEF and Avangard.[33] Accordingly, the management of the institute began to turn to city authorities for requests for loans and extra-budgetary funding for specific projects. Most importantly, VNIIEF and Avangard needed to reschedule their municipal tax payments.[34] Without these extensions, the enterprise would have incurred additional fines that, in turn, would have exacerbated the debt problem.

In addition to low-interest loans and the rescheduling of tax payments, the Sarov administration assisted VNIIEF in supporting the institute's infrastructure. In particular, the municipal budget financed specific projects and security upgrades at the institute. Consequently, by 2000, VNIIEF had received approximately 500

---

[31] Radiy Ilkayev made this note in 1998, when relations between the city and institute had already changed significantly, which goes to show the strength and resilience of the decades-long tradition of supremacy of the institute director in all spheres of city life.

[32] This process was accompanied by the growth in prosperity of the city authorities. City administration buildings were refurbished and new cars were purchased for city bureaucrats. Bonuses were given, loans, good housing, etc. were offered. All these signs of power began to appear quickly, and the city gained this new social and material well-being not from VNIIEF or Avangard, like in the past, but on its own—from the investment zone.

[33] Attempts to describe the relationship between the institute and the city in terms of "richer" or "poorer" are somewhat incorrect. Even after the creation of the investment zone, VNIIEF's budget was greater than the city's. In 2000, Sarov had a budget of 1.7 billion rubles (at the year's average exchange rate, slightly over $60 million). The institute's budget, according to our estimates, totals about $100 million per year. Naturally, one must take into account that the institute pays part of this money to the city as taxes. More important than the size of the annual budget is that the financial position of city authorities was more stable, with no late tax payments (other than those of VNIIEF and Avangard). The city's receipt of taxes from other firms and enterprises changed the economic and everyday relations between the city and enterprise.

[34] The city authorities had the right to grant extensions until the end of the fiscal year, i.e., December 31 of the current year.

million rubles in assistance from the city administration.[35] These loans and assistance effectively rendered the institute indebted to the city administration—a situation that was unprecedented and also very difficult for the leadership of VNIIEF to accept.

VNIIEF and Nizhniy Novgorod Oblast

In 1997, then-Nizhniy Novgorod Governor Nemtsov attempted to interfere in the process of appointing a new director at VNIIEF. In late 1996, Nemtsov's protégé Krysov had been appointed to the post of director of VNIIEF-Konversiya and simultaneously to the post of deputy director of VNIIEF. From all appearances, Nemtsov was lobbying on behalf of Krysov in the hopes of gaining an ally to spearhead conversion at the institute. However, because the governor could not formally participate in the process of appointing a new VNIIEF director, the issue was decided in Moscow, with input from Minatom. Minister of Atomic Energy Mikhailov staunchly supported Ilkayev for the post of director. Ilkayev was a member of the "weaponeers" group that traditionally managed the institute. Mikhailov ultimately prevailed and thwarted Nemtsov's initiative, which, in turn, effectively soured relations between Minatom and the governor. In the summer of 1997, Nemtsov was appointed deputy prime minister of Russia. His successor as governor of Nizhniy Novgorod Oblast was Ivan Sklyarov, who received approximately 52 percent of the popular vote in the region.[36] As governor, Sklyarov spoke out very favorably regarding the management at VNIIEF. Sklyarov also played a critical role in assisting with debt relief for VNIIEF and the military-industrial complex across the oblast. During a period of acute wage arrears suffered by VNIIEF, Sklyarov lobbied the federal government for assistance. Together with the Ministries of Finance and Economy, the oblast administration also hammered out an agreement in 1998 for settling the wage arrears of defense industrial enterprises throughout Nizhniy Novgorod Oblast.[37]

On the whole, the new governor established good working relations with the VNIIEF leadership. In 1998, it was rumored that Ilkayev would run for office in the Nizhniy Novgorod Oblast Legislative Assembly. In general, VNIIEF worked to increase the prestige of Nizhniy Novgorod Oblast in return for the financial assistance provided by the local authorities.[38] However, VNIIEF did not receive

[35] Part of this figure consists of pre-crisis rubles (the average exchange rate at the end of 2001 approached 30 rubles to the dollar, while prior to the 1998 crisis, there were 6 rubles to the dollar).

[36] Sklyarov's opponent was Gennadiy Khodyrev, who received 42.15 percent of the vote. Sklyarov, however, received a lower percentage of the vote in Sarov than in other cities of the region. He held office for one term, ending in 2001.

[37] At that moment, the oblast's MIC enterprise debts were 460 billion rubles, with wage arrears accounting for 137 billion.

[38] For example, VNIIEF's appearances at international expositions were sometimes sponsored by the oblast budget. VNIIEF Chief Engineer A. Kovtun noted this in a 1998

significant infusions from the oblast budget. The governor and his circle asserted that Nizhniy Novgorod Oblast was hardly in the position to provide relief for VNIIEF, as the institute's employees were considerably better off than the average oblast resident. Instead, the oblast leadership wanted the regional administration and the whole region to profit more from the operation of the investment zone in Sarov.[39]

Sarov and Nizhniy Novgorod Oblast

Relations between the city and the oblast have been marred by friction since the Soviet collapse. A point of contention was the redistribution of federal funding allocated to the oblast for public health following the aftermath of the Chernobyl disaster. In 1997, for example, the region did not pass on any of the federal allotments to Sarov. Sklyarov subsequently claimed both that the federal government was delinquent in disbursing the funds and that those monies received were used to construct new apartment buildings in another town that would house Chernobyl victims.[40] He also did not consider it necessary to "share" regional funds with Sarov. This echoed a sentiment shared among regional administrators, who generally assumed that Sarov continued to receive preferential treatment from the federal government, and that after the creation of the investment zone the city was attracting taxpayers who would otherwise be paying into the oblast budget.[41] A similar lack of sympathy for the problems confronted by VNIIEF and Sarov pervaded within the Nizhniy Novgorod Oblast Legislative Assembly.[42]

In the fall of 1997, Governor Sklyarov visited Sarov to discuss the operation of the investment zone. Regional authorities were alarmed by the outflow of

---

interview. It may be that he was only talking about instances when VNIIEF was represented at the oblast's booth at certain expositions, rather than all occasions when VNIIEF participated in international trade shows.

[39] Conversation with O. A. Kolobov, head of the department of international affairs at Lobachevsky Nizhniy Novgorod University and advisor to Governor Sklyarov, Nizhniy Novgorod, January 19, 2001.

[40] *Gorodskoy Kuryer*, No. 52, 1998.

[41] City media sources described the governor's interest thus: "The governor's interest in his most recent visit was targeted at zone business." Not every interested company was registered in the zone. A stringent selection process was followed. The formal goal of the selection process was to choose the investment and conversion projects most attractive for the city. There seemed to be another, informal, goal: to sift out small firms. Since the City Duma and local authorities viewed the zone as a temporary arrangement from the very start (initially for two years), they limited the number of enterprises that could register to 200. There was no limit on the amount of tax privileges at that time, so the idea was to collect as much as possible from these 200, which predisposed the city toward attracting mainly large taxpayers.

[42] Conversation with O. A. Kolobov, head of the department of international affairs at Lobachevsky Nizhniy Novgorod University and advisor to Governor Sklyarov, Nizhniy Novgorod, January 19, 2001.

taxpayers, including major ones, from the region. The new governor managed to make some critical (yet guarded) comments concerning the investment zone. The climax of this visit was the signing, by Sklyarov and the mayor, of a protocol consisting of five points: 1) the experience with the zone was dubbed positive; 2) VNIIEF-Konversiya was obliged to provide the governor with monthly reports of its progress; 3) by the end of 1997, 10 additional Nizhniy Novgorod enterprises were to be registered in the zone; 4) a customs post was to be established in the city; and 5) a specialized investment fund for Nizhniy Novgorod Oblast was to be created.

But the final point of the protocol was the most interesting. It was apparently the main compromise between regional and local authorities. The money from this fund was to be used for financing joint projects between Sarov and Nizhniy Novgorod Oblast. These projects were to involve high technology and would be implemented either wholly or partially outside Sarov. On December 18, 1998, Nizhniy Novgorod authorities and Sarov signed an agreement on mutually beneficial cooperation in using the results of investment zone activities.[43] The parties agreed to coordinate their activities and foster the creation of jobs and industrial capacity. The Sarov administration, in turn, agreed to attract new participants in the investment zone only with the consent of the Nizhniy Novgorod administration.

Thus, at first, Governor Sklyarov was critical of the Sarov investment zone and its effects on lowering oblast budget revenues. But, gradually, the governor came around to support the zone, albeit as a temporary arrangement.[44]

It is interesting to note, that in the 2001 gubernatorial election, Sklyarov lost to a "moderate" Communist candidate Gennady Khodorev, garnering fewer votes from Sarov. In fact, voter turnout in Sarov was 28.9 percent, down significantly from 38.8 percent in the 1997 election.[45] This could be explained by the growing apathy and pessimism among Sarov residents who blamed Sklyarov for the perceived deterioration in their personal financial situations. Yet, in fact, economic conditions in Sarov seemed to improve during Sklyarov's tenure. This upswing was largely attributed to the city's special status as an investment zone. Accordingly, Sarov residents had became more attuned to the policies pursued by

---

[43] This agreement was signed with a duration of one year. Unfortunately, there is no data indicating how it was implemented or whether it was extended.

[44] As Sklyarov explained, "At the outset, I said that if there are no economic actors that come to your zone and create jobs...there's nothing to talk about. But later I started to count. The oblast budget received R97 billion less, because 71 of our enterprises are registered in your zone. I consider these temporary losses, since the enterprises are beginning to flourish and work. The Sarov budget saw a profit of approximately the same amount: R70 billion plus R40 billion to the investment funds. Our loss is R100 billion, your gain is R100 billion—this is the money we're talking about. But the enterprises themselves received R200 billion in investment, which allowed them to get on their feet. Our mistake is in not being demanding enough when it comes to the business plans." *Gorodskoy Kuryer*, No. 51, 1998.

[45] *Gorodskoy Kuryer*, No. 210, 2001.

Minatom or the Ministry of Finance in Moscow, as well as by municipal authorities, as opposed to those championed at the oblast level.

Sarov: Observations on the Decade from 1991 to 2001

Notwithstanding the objective hardships experienced at Avangard, VNIIEF, and Sarov, conditions in the city throughout the 1990s remained stable relative to the all-Russian average. Even during the operation of the investment zone with the infusions of additional revenue streams, the city continued to receive significant subsidies from Moscow. That said, the city experienced considerable change over the decade. In particular, the establishment of the investment zone put the city on the map, both literally and figuratively. On the one hand, the city's new status allowed new business activities to flourish that were unrelated to VNIIEF or Avangard. On the other hand, the opening introduced new problems.

By 1996 and 1997, the city began to experience a significant rise in the levels of drug use (primarily heroin). Since then, drug abuse has become a major problem for city authorities. The Sarov administration adopted a special program (with targeted financing from the city budget) to prevent the spread of drugs. Similarly, beginning in 1997-98, the number of burglaries in the city began to rise sharply. Law enforcement officials and the mayor's office attributed this to the rise in drug use. Despite stepped-up security measures—including checkpoints and an outer perimeter, restrictions on entering the city, a pass system, and background checks on all persons entering the city—local law enforcement officials have been relatively unsuccessful at stemming the smuggling of drugs and stolen goods into and out of the city.[46] Not only are there petty thieves in the city, but organized criminal groups—large mafia organizations and professional criminals ("thieves in law")—have begun to prey on the city, as well. Sarov, for example, has its own so-called city "observer" (or "curator" in the lingo of the criminals), who watches out for the interests of organized criminal groups.[47]

These emerging conditions in the city have spawned a review of traditional security practices within Sarov. Under the Soviet system, the focus was on monitoring the human factor. This method was rendered obsolete with the range of socio-economic changes and growing respect for human rights in the "new" Russia. Instead, attention has been increasingly paid to improving perimeter

[46] Virtually all stolen property is sold outside of the city. Selling second-hand goods in Sarov is too noticeable.

[47] Conversation with high-ranking official in Sarov city administration, Sarov, January 18, 2001. At present, however, there is no sign that organized crime gained access to the finances of city enterprises or the city budget. The upsurge of crime has agitated the institute's security services. Approximately 3,000 ex-convicts live within the ZATO. For a city with a population of 80,000 that is used to a high level of security, this is an extremely high number. There is widespread concern among local security personnel that terrorists could enlist such segments of the local population as accomplices that could readily facilitate access to the city and increase the transfer of illicit materials.

defense. In 1997-98, 83 billion rubles were allocated from investment zone income for security programs at VNIIEF. These funds were used to reconstruct perimeter fences and renovate the barracks of security personnel. However, the funding was insufficient to complete the planned measures to bolster the level of security at VNIIEF."[48]

At the same time, elements of civil society have begun to take root in the city. Sarov has its own media and a number of non-governmental organizations that were unthinkable for a closed city under the Soviet regime. Sarov even has a non-governmental organization that focuses on nonproliferation issues.[49]

The oversized workforce at VNIIEF and Avangard has been another serious problem for the city. Minatom charged VNIIEF with reducing its workforce by one-third to 12,000 employees. According to VNIIEF and Minatom, this would be sufficient to fulfill state defense orders but would nonetheless impose a significant social burden on the city, as at least 6,000 VNIIEF employees would be released.[50] To alleviate possible aggravation of the social situation in the city the emphasis was placed on the creation of alternative civilian jobs and business development in Sarov. Besides possible social problems, it was clear that letting nuclear specialists go without meaningful alternative employment could create a potential nonproliferation problem. In December 1998, for example, a VNIIEF employee was arrested for an attempted sale of non-nuclear weapon technical documents to reportedly Iran or Afghanistan.[51]

International nonproliferation assistance programs have played a vital role in containing the brain-drain from VNIIEF. Particularly, this assistance was crucial during the mid- and late 1990s. The assistance was provided through the International Science and Technology Center's research grants to VNIIEF scientists and through contracts under the U.S. government laboratory-to-laboratory and later Material Protection, Control, and Accounting (MPC&A) program.[52]

---

[48] Conversation with high-ranking employee of VNIIEF responsible for security issues, Sarov, January 18, 2001. In the future, the absence of additional income from the investment zone will make security upgrades increasingly difficult to complete. Although the federal budget does contain an item on increasing financing, which has become more reliable in connection with Moscow's policy to increase physical protection of nuclear facilities, which was activated after September 2001 in order to prevent nuclear terrorism. Interview by FSB chief N. Patrushev to *Gorodskoy Kuryer*, No. 235, 2001.

[49] Analytical Center for Nonproliferation in Sarov, <http://www.npc.sarov.ru>.

[50] Ye. Mazanova, "Na press-konferentsii, posvyashchennoy itogam raboty....," *Gorodskoy kurer* online edition, January 4, 2001, <http://infra.sar.nnov.ru/~courier/>.

[51] NTV, December 18, 1999; in "Nuclear Centre Worker Caught Selling Secrets," Lexis-Nexis Academic Universe, <http://www.lexis-nexis.com/universe>.

[52] Some VNIIEF employees involved in these programs earned up to $800–$2,000 per month, though typically a mid-level special was paid $400 while working on an ISTC project. This is a very hefty salary in the provinces. The institute usually had two to three thousand involved in these projects, and the individuals did not remain the same, but were

The Sarov investment zone was sought as one of the powerful means to boost the city economy and create new civilian jobs. However, the "offshore" privileges were terminated in 2001. Debate surrounding the appropriateness of the zone turned on assessments of the utility of subsidies. On the one hand, the proponents of the zone remained skeptical of the government's commitment to Sarov. They were convinced that: 1) given current circumstances, centralized financing would never be disbursed fully or regularly; and 2) the implementation of conversion programs required private investment that, in turn, depended on financial incentives and less intrusive state intervention.[53] Alternatively, critics of the zone stressed the benefits of government subsidies. State support offered greater order, since subsidies are disbursed in a targeted manner. Also, it provided a mechanism to insulate business activity from corrupt and overzealous interference by regional and local authorities.

Another U.S. nonproliferation program—the Nuclear Cities Initiative (NCI)—has also assisted in the creation of new jobs and businesses in Sarov. Sarov has received the lion's share of the NCI funds, particularly in the first years of NCI. Several viable and profitable businesses have been developed with the NCI funding—such as the Open Computing Center, a Laparoscopy Center, and several other projects. At the same time, the overall number of new jobs created with NCI funding over the course of five years did not exceed several hundred.[54]

By the end of the decade, VNIIEF's financial situation began to stabilize. Payments for state defense orders arrived regularly after 1998. Despite the qualitative and quantitative changes in the city, VNIIEF remained an enormous institute, oriented primarily toward the military-industrial complex. Much of the stability could be attributed to international assistance, which accounted for 7 percent of the institute's budget in 1997 and grew to 13 percent in 2000. Yet, the institute management had not intended to see the percentage of international assistance grow and was determined to maintain VNIIEF's self-sufficiency and independence. While it was amenable to expanding cooperation with private Western companies on a purely commercial basis, the management of VNIIEF remained wary of expanding participation in programs supported by foreign governments.

Under President Putin, the newly appointed presidential representative (*polpred*) to the Volga Federal Okrug, Sergey Kiriyenko, the former Russian prime minister and one of the leaders of the Union of Right Forces, has assumed a relatively low profile on issues related to VNIIEF and life in Sarov. For the most

---

rotated through the projects (in order to allow the largest number of employees to earn a little extra).

[53] Conversation with deputy of the Sarov City Duma, Sarov, January 17, 2001.

[54] Jean Pierre Contzen and Maurizio Martellini, "Russian Nuclear Weapons Complex Workforce Downsizing Through Secure Retirement Rationalities and Proposals," September 2003, <http://www.sgpproject.org/publications/Contzen&Martellini.html>.

part, the *polpred* has confined his role to articulating federal policies[55] and presenting government awards for achievement to personnel at VNIIEF. Yet, this diffidence was the result of interests and distractions peculiar to Kiriyenko, rather than a reflection of concerted federal policy. In particular, Kiriyenko also was appointed by the president to head the Commission on Chemical Weapons Elimination. His primary responsibility, in this regard, has been to seek out extra-budgetary finances for chemical weapons elimination in the okrug and to promote Russia's chemical demilitarization efforts as a whole. Kiriyenko's involvement in Sarov's issues has become more active with the increased attention to nuclear security after the terrorist attacks on the United States in 2001 and with the preparations to the celebration of St. Seraphim's anniversary in 2003, which entailed enhanced security in Sarov and around its nuclear facility during the celebration—when the city opened its gates to a large number of religious worshipers.

Putin Reforms and Recent Trends

Putin's federal reforms put an end to separatist tendencies in Russia and established much stricter control over the federation's constituencies, particularly in the area of federal jurisdiction. Coupled with the stabilization and growth of Russian economy, these changes have translated into increased and more stable funding for Sarov and VNIIEF. By the end of 2002, the average salary in Sarov was 1.5 times higher than in Nizhniy Novgorod.[56] Moreover, salaries at VNIIEF are much higher than average city salary, and wages are paid on time, while the state defense order is steadily increasing.[57] In December 2002, an average salary in science sector was 12,406 rubles (approximately $400) a month. This stabilization has reinforced VNIIEF reliance on Minatom and federal government. However, it does not mean that the relationships with the oblast and, particularly the city, have returned to those of the Soviet period. Moreover, 2003-04 changes in federal legislation defining municipalities could mean a return to a much more active reliance on local and regional resources in providing for the laboratory and the city. For example, in January 2004, VNIIEF asked the city to provide the institute with a 90 percent relief from its property tax, which was not calculated in the

[55] Nizhniy Novgorod Oblast has fared well in bringing regional legislation into line with federal laws. The few inconsistencies that have occurred are not the result of targeted intent. In some instances, regional legislators simply have not managed to react quickly enough to changes in federal law. There are some examples of federal okrug authorities acknowledging that local laws are "better" than federal laws and presenting legislative initiatives to change federal law. Moreover, there has not been a single discrepancy between the legislation of Nizhniy Novgorod Oblast, federation constituencies in the okrug, and national legislation concerning national security, defense, and nuclear issues.

[56] L. Saratova, "Istina–v sravnenii," *Gogodskoy Kuryer*, No. 30, 2003.

[57] Olga Zaguskina, "Kak deputaty budut otvechat pered izbiratelyami," *Gorodskoy Kuryer*, No. 2, January 8, 2004.

federal contribution but which the institute was required to pay to the city according to new legislation. The city budget could encounter a much worse economic loss if the closed cities lose their direct federal funding and become subordinated to the oblast in 2005.[58]

Another interesting development, which could also be attributed to the changes that occurred in 2000-03, is a sudden change in the fate of the Avangard plant and its workforce. The downsizing plans of the Russian nuclear warhead production complex, at least until recently, provided for a complete termination of nuclear defense work at Avangard by 2003 and conversion of its infrastructure to civilian production or least non-nuclear-weapon work.[59] It was expected that former Avangard employees would either be engaged in new conversion projects at the plant or absorbed by newly created businesses in Sarov. In mid-2003, the Avangard Electromechanical Plant instead was merged with VNIIEF, including the merger of some of its infrastructure and employees.[60] Staff reductions at VNIIEF have not materialized either. To the contrary, by January 2004, VNIIEF employed 25,000 employees including a hefty addition from Avangard.[61] How long Minatom and VNIIEF will be able to sustain this workforce is unclear. VNIIEF Director Ilkayev has already voiced his dissatisfaction with Minatom's support in implementing the merger.[62]

The failed reduction of VNIIEF staff, including those involved in defense work, is even more disappointing, because the total investments into the conversion projects in Sarov were the largest ones among Minatom's nuclear closed cities. Sarov is the champion recipient of both Minatom funds and foreign assistance money for conversion projects and job creation programs. Minatom praised Sarov as the most successful among closed cities in implementing conversion and non-weapon programs.[63] But the oversized workforce at VNIIEF remains a major drain on government financial resources, while staff reductions continue to be on the agenda of the institute and Minatom.

---

[58] Ibid.

[59] Novosti OPK i VTS," *Nezavisimoye voennoye obozreniye* on-line edition, May 11, 2001, <http://nvo.ng.ru>.

[60] "Spustya 46 let zavod 'Avangard' (Sarov) stal strukturnym podrazdeleniyem Vserossiyskogo NII Eksperimentalnoy fiziki," Privolzhye information agency; in Integrum Techno, June 16, 2003, <http://www.integrum.com>.

[61] Olga Zaguskina, "Kak deputaty budut otvechat pered izbiratelyami," *Gorodskoy Kuryer*, No. 2, January 8, 2004.

[62] Elena Mazanova, "Deneg by pobolshe," *Gorodskoy Kuryer*, No. 2, January 8, 2003.

[63] "Minatom nazval yadernyy tsentr v Sarove odnim iz luchshikh predpriyatiy po osvoyeniyu neyadernykh vooruzheniy," Privolzhye Information Agency, January 22, 2003; in Integrum Techno, <http://www.integrum.com>.

Conclusion

With the exception of the Zemlya movement's short-lived attempt to claim the territory of Sarov, there was no serious direct or indirect attempt by regional or local authorities to challenge federal control over VNIIEF during the 1990s. In general, all of the sub-federal levels of government shied away from assuming the complicated and difficult tasks of supervising the institute. Rather, authorities at each level sought to distance themselves from responsibilities of oversight and control and to scale back interaction with the institute. The federal government continuously withheld subsidies and oblast officials were loath to step in with offers of substantial financial relief. Even when the Sarov administration received money from the investment zone, it too was reluctant to "share" with the institute and was careful to limit the scope of its support. At all levels of government, VNIIEF was regarded as a drain on resources, as opposed to a potential donor. Consequently, there was very little interest in exploring alternative mechanisms for improving VNIIEF's "bottom line."

In general, the difficult socio-economic situation confronting VNIIEF in the 1990s and the institute's low political profile insulated it from being a subject of center-periphery disagreement. The most critical change in relations with the oblast and local administrations was attributable to the creation of the investment zone, which affected relations between VNIIEF and the Sarov administration the most. While the institute had initially supported the law on the closed cities, both the law and related amendments granted tax privileges that effectively lined city coffers. At first, this caught VNIIEF by surprise, and the management presumed that the creation of the investment zone would provide the institute with "financial independence" from the center and insulate it from the difficulties of obtaining state defense orders. When it became clear that this additional income would not come under VNIIEF's control, Director Ilkayev began to lobby Moscow for tax exemptions, while continuing to receive traditional government financing and additional payments for conversion programs and international assistance. There was widespread hope within the institute that this would significantly improve VNIIEF's financial position and increase its independence.[64] Instead, the investment zone created an independent city administration and legislation that no longer rubber-stamped decisions made by VNIIEF or Minatom. The investment zone and conversion efforts laid the foundation for economic development in Sarov. However, these efforts did not succeed in ensuring the city's self-sufficiency. As of 2004, Sarov continued to receive subsidies from the federal budget.

Overall, Sarov was not integrated into the economic and political sphere of Nizhniy Novgorod Oblast either. Oblast authorities considered the city and its enterprises generally to be a drain on its resources. Since the oblast itself was suffering constantly from budget deficits in the 1990s, it did not want to deal with

---

[64] The institute management also actively sought lobbyists in Moscow that could protect VNIIEF's interests. One of the institute's allies was even the Russian Orthodox Church.

another recipient. Of equal, or even greater, importance is that Sarov was simply of little interest from a regional political standpoint. The oblast has a population of 3.6 million.[65] Sarov's population of 84,000 is an insignificant percentage of the electorate. Ironically, Sarov is more interesting in the national political arena, particularly for politicians who want to use national security issues for their political gain.[66]

While the political and economic situation in closed cities and around them stabilized in 2000-03, no serious structural changes in the operation of nuclear facilities or the city itself have taken place. The downsizing of VNIIEF and financial and economic self-sufficiency of the city have still not been achieved. The planned "re-subordination" of Sarov from the federal government to Nizhniy Novgorod Oblast could pose a new challenge to the city's well-being and potentially create grounds for some heated confrontation with oblast. In the end, however, these changes will increase reliance on local resources and speed up the integration of Sarov into the oblast and the broader Russian economy.

---

[65] As of January 1, 1999, 3,687,700. *Regiony Rossii 1999* (Moscow: 1999), p. 167.

[66] It seems that this is the reason that in the mid-1990s practically all well-known politicians visited Sarov, some more than once. The city authorities and the institute tried to use the politicians' desires to visit Sarov to expand their opportunities for lobbying.

Chapter 8

Nuclear Issues in the Urals Federal Okrug

Elena Sokova

Most of the nuclear weapons production facilities in the Urals were constructed from the late 1940s through the 1950s. By the 1980s, the Urals nuclear complex reached its peak in terms of size and output. However, the end of the Cold War and Russia's economic hardships following the collapse of the Soviet Union in late 1991 profoundly affected nuclear enterprises in the Urals. Many of them had to precipitously scale back defense production significantly. They had to convert at least partially, as was the case for the serial production facilities in Trekhgornyy and Lesnoy, or completely, as at the Urals Electrochemical Combine in Novouralsk, where activities shifted from the production of highly enriched uranium (HEU) to its downblending for low-enriched uranium (LEU) production.

Yet, even drastic cuts in defense orders and downsizing efforts in the nuclear weapons complex did not lead to the closure of any of Minatom's nuclear enterprises in the Urals. Moreover, according to a May 7, 2001 government decree (*rasporyazheniye*) on restructuring the defense industry for the period until 2010, all nuclear warhead manufacturing in Russia was slated to be concentrated in facilities in the Urals. The Mayak Production Association in Ozersk, for example, was designated as the sole Russian site for manufacturing fissile material components for nuclear warheads. Two production facilities that will continue assembly and disassembly of nuclear warheads are also in the Urals—the Elektrokhimpribor Combine in Lesnoy (Sverdlovsk-45, Sverdlovsk Oblast) and Instrument-Making Plant in Trekhgornyy (Zlatoust-36, Chelyabinsk Oblast).[1]

Notwithstanding the administrative changes, regionally based military units and elements of the Minatom nuclear warhead production complex remained federal property, subject to direct federal control.[2] Military units have remained directly subordinate to Moscow, and there have been no reports of any political opposition

[1] Novosti OPK i VTS, *Nezavisimoye voennoye obozreniye* online version, May 11, 2000, <http://nvo.ng.ru>.

[2] Several Minatom enterprises that are not involved in the production of nuclear weapons or materials, such as SverdNIIkhimmash, Malyshevskoye ore directorate, and the Nizhyaya Tura Venta Machine-Building Plant were transformed into joint-stock companies, but the government still holds about 50 percent of their shares.

to the federal center or suspicious ties to regional or local governments. Local and oblast administrations have provided some basic necessities, such as food and supplies, and helped with housing construction and income-generating arrangements, but have refrained from more active involvement. In addition, there have been only *ad hoc*, individual attempts to establish a "special relationship" with the military, such as the 1998 agreement between the political organization "For the Rebirth of the Urals" (*Za vozrozhdeniye Urala*) and the special airborne brigade in Asbest-5, Sverdlovsk Oblast. Moreover, these incidents typically provoked negative responses from the regions and were immediately abandoned.[3]

Minatom facilities in the Urals, in contrast to Ministry of Defense (MOD) facilities, have a wide range of interaction with regional and local authorities in the economic, social, environmental, safety, and security areas. Nuclear facilities and nuclear policy have often become a focal point of regional politics. In turn, local and regional factors continue to play a significant role in shaping the environment in which nuclear facilities operate.

After the breakup of the Soviet system, nuclear facilities, all of which were previously oriented exclusively toward Moscow, had to learn that on certain issues—from energy supply to their cities and facilities to the appointment of nuclear facility directors—they now have to communicate and coordinate their actions not only with Minatom (or its predecessor, Minsredmash), but also with regional and local governments. As summed up by the General Director of Mayak, Vitaliy Sadovnikov:

> Mayak cannot live in isolation from the city, oblast, and the country.... [...] The time when Sredmash was a state within a state is over.... [...] Mayak should find its place within the city, oblast, and country, including the political and information landscape.... [...] We care who the president of our country is, who the oblast governor is, who represents Ozersk in the oblast Duma or city council, or heads the city administration.... [...] Consequently, we will try to influence these processes actively.[4]

This chapter focuses on key areas of interaction between regional and local governments and the nuclear facilities in the Urals subordinated to Minatom. Special attention is devoted to detailing how these interactions play out along political, economic, and social dimensions in the Chelyabinsk and Sverdlovsk Oblasts. Finally, there is a discussion of how this interplay affects nuclear facilities and the formation and implementation of nuclear policy.

Before proceeding, however, it is important to note the significance of the Urals' footprint in the Russian nuclear complex. The Urals Federal Okrug (district) consists of six regions: Chelyabinsk, Kurgan, Sverdlovsk, and Tyumen Oblasts, and Khanty-Mansiysk and Yamalo-Nenetsk Autonomous Okrugs. Of the members of the district, only Chelyabinsk and Sverdlovsk Oblasts have nuclear facilities and

---

[3] Aleksandr Krymov, "Zachem 'ZVU' svoy spetsnaz?" *Chelyabinskiy rabochiy*, June 17, 1998; in National Electronic Library, <http://nel.nns.ru>.

[4] Elena Vyatkina, "'Mayak' vo vremeni i prostranstve," Ozersk website, September 6, 2000, <http://www.ozersk.ru>.

thus are the focus of this chapter. Five out of Minatom's 10 nuclear closed cities are located here. Other nuclear infrastructure includes two ICBM bases, several warhead and missile storage facilities, strategic nuclear weapons production facilities, a number of nuclear research institutes, and a nuclear power plant (see Tables 8.1 and 8.2).

Table 8.1 Nuclear Infrastructure: Chelyabinsk Oblast

Location	Name of Facility	Description and Comments
Kartaly or Lokomotivnyy MOD closed city	ICBM base	46 RS-20 (SS-18) ICBMs
Snezhinsk *(formerly Chelyabinsk-70)* Minatom closed city	All-Russian Scientific Research Institute of Technical Physics (VNIITF)	Nuclear weapons research and design lab
Trekhgornyy *(formerly Zlatoust-36)* Minatom closed city	Instrument-Making Plant	Nuclear warhead assembly, dismantlement, and storage; production of ballistic missile reentry vehicles
Ozersk *(Chelyabinsk-65, also known as Chelyabinsk-40)* Minatom closed city	Mayak Production Association (PO Mayak)	Fissile material component production; storage of HEU, civilian and weapons-grade Pu; tritium production reactors; spent fuel reprocessing; nuclear waste vitrification; storage facility for fissile material from dismantled warheads, experimental MOX fuel production facility
	All-Russian Scientific Research and Design Institute of Power Engineering Technology (VNIPIET) Urals branch	VNIPIET (headquarters in St. Petersburg) specializes in the design of nuclear reactors
Miass	Makeyev Design Bureau	SLBM design bureau
Zlatoust	Zlatoust Machine-Building Plant	SLBM components manufacturing and SLBM assembly

Table 8.2 Nuclear Infrastructure: Sverdlovsk Oblast

Location	Name of Unit	Description
Lesnoy *(Sverdlovsk-45)* Minatom closed city	Elektrokhimpribor Combine	Nuclear warhead assembly, dismantlement, and storage; isotope production
Malyshev, near Asbest	Malyshev Mining Utility/Ore Directorate	Mining of uranium ore using the in-situ leaching method in the Zauralskiy ore region, Kurgan Oblast. In August 2001 became a part of TVEL
Nizhnyaya Tura	Nizhnyaya Tura Venta Machine-Building Plant	Production of weapons parts and technology and equipment for reprocessing and storing nuclear waste
Novouralsk *(Sverdlovsk-44, also known as Verkh-Neyvinsk or Yekaterinburg-44)* Minatom closed city	Urals Electrochemical Combine	Blending down of weapons-grade uranium into low-enriched uranium; uranium enrichment for nuclear fuel. Storage of weapons-grade uranium
Svobodnyy *(Nizhniy Tagil)* MOD closed settlement	ICBM base	45 RS-12M (SS-25) ICBMs (as of July 2000)
Uralskiy MOD closed settlement	ICBM storage facility	Storage facility for ICBMs without nuclear warheads
Yekaterinburg	"SverdNIIkhimmash" Joint Stock Company	Design bureau for radioactive waste processing equipment, including the deactivation and reprocessing of solid metal waste from decommissioned nuclear power reactors and nuclear weapons
	Urals Electrical and Mechanical Plant	Non-nuclear warhead components production

Table 8.2 Nuclear Infrastructure: Sverdlovsk Oblast (continued)

Location	Name of Unit	Description
Zarechnyy Partially closed city subordinated to Sverdlovsk Oblast	Beloyarsk Nuclear Power Plant	Unit 1: AMB-100 water-graphite channel-type reactor (shut down) Unit 2: AMB-200 water-graphite channel-type reactor (shut down) Unit 3: BN-600 sodium-cooled fast-breeder reactor (operational) Unit 4: BN-800 (under construction)
	Scientific Research and Design Institute of Energy Technologies (NIKIET), Sverdlovsk branch	NIKIET designs and develops nuclear reactors, including graphite-moderated reactors for power generation, for propulsion and power generation in outer-space vehicles, for naval propulsion, for heat production, and for research. It also conducts research in the areas of reactor materials and reactor physics, and develops and tests instruments and control systems for the nuclear power industry

Political, Economic, and Social Conditions in Chelyabinsk and Sverdlovsk Oblasts

The Chelyabinsk and Sverdlovsk Oblasts rank relatively high among other Russian regions in economic terms.[5] The two oblasts are major producers of weapons, both nuclear and conventional. However, natural resources, metallurgy, machine building, and the chemical industry have been and remain major sources of revenue and account for the bulk of industrial output and exports in both oblasts. The population in Chelyabinsk and Sverdlovsk Oblasts is predominantly Russian: 88.7 percent in Sverdlovsk Oblast and 81 percent in Chelyabinsk Oblast.[6] To date, there have been no significant problems of an ethnic, religious, or separatist nature.

In spite of these marked similarities, Sverdlovsk and Chelyabinsk Oblasts differ in important respects. The most conspicuous differences pertain to the ability of the respective regional governments to lobby Moscow on behalf of the oblasts' interests, the influence of national financial-industrial groups (FIGs) on their territories, and respective relations with Minatom. The two oblasts compete with each other in many spheres, from metallurgy to state military orders. In most areas, Sverdlovsk Oblast is in the lead, owing primarily to its relative economic

[5] Sverdlovsk Oblast ranks six among Russia's regions, while Chelyabinsk Oblast takes the fourteenth place. Investment Ratings of Russia's Regions. Ministry of Economic Development and Trade of the RF, <http://www.economy.gov.ru>.

[6] Handbook on Russian Regions, EastWest Institute Regional database, <http://www.iews.org>.

performance and the political clout its governor, Eduard Rossel enjoyed in the late 1990s.

In the early and mid 1990s, Sverdlovsk Oblast, led by its governor, was also at the forefront of the movement for regional autonomy, although these aspirations were of an economic, rather than political, nature. Governor Rossel championed greater autonomy for Sverdlovsk Oblast, as well as advocating an upgrade of the region's status to that of a republic, thus allowing for more economic freedom and a greater share of taxes going to regional coffers. Rossel also spearheaded movements to establish the Greater Urals Economic Association and the Urals Republic that existed for only 10 days before it was disbanded by President Boris Yeltsin. The former gained considerable influence in the region and could have posed a serious challenge to federal policies there. The introduction of federal districts in May 2000, which divided members of the association between the Urals and Volga districts, as well as the new political landscape in federal relations in Russia, considerably reduced the political clout of the Greater Urals Economic Association.

Governor Rossel also was credited with promoting the election of governors by popular vote. He became Russia's first elected, rather than appointed, governor in August 1995. Until 2002, Rossel continued to be the most vocal advocate among heads of non-ethnic regions for "real federalism": a clear division of powers between the center and the regions and greater independence in the economic sphere and for "fair" distribution of taxes between Moscow and regions. This put him on a collision course with federal authorities in 2000-2002. He often was at odds with Urals District Presidential Envoy Petr Latyshev. By mid-2002, however, it became clear that the federal center had won this tug of war, and that Rossel had to tone down his criticism of the Urals' presidential envoy and his anti-Moscow rhetoric, particularly when Moscow threatened to withhold support for his reelection campaign in 2003.

Before Putin initiated federal reforms in 2000, Rossel was one of the most influential governors and a prominent political figure at the national level. He used his political clout to lobby Moscow on behalf of the oblast's defense sector, and to promote conversion at many of these facilities. His membership in the Federation Council's Committee on Security and Defense in the mid-1990s, as well as his personal participation in senior delegations to foreign countries helped him to succeed in many of his efforts. Rossel actively promoted greater independence for conventional weapons producers in their trade with foreign customers, arguing that the centralized weapons sales system had led to a loss of potential customers and contracts.[7]

The motive behind this "pro-independence" push was mostly economic. In particular, Rossel sought to keep defense enterprises afloat at a time of rapidly shrinking state defense contracts. Of course, one has to be sensitive to the possible

---

[7] The centralized system decreases the amount from sales that goes directly to the enterprises and, thus, decreases tax revenues for the oblast. In addition, the oblast bank system is stripped of a significant inflow of cash.

proliferation impact of "free exports" upon international security, but these concerns were not high on Rossel's agenda. In 1998, Rossel visited Iran, offering equipment for steel mills, deliveries of metal products, and nuclear fuel supplies.[8] Although his "initiative" did not generate new contracts in the nuclear sphere, it prompted federal authorities to launch a series of measures to ensure that Russian regional administrations would operate strictly within the limits of the powers defined in the Constitution and other federal legislation on issues related to foreign trade.[9]

Chelyabinsk Oblast, though it followed Sverdlovsk Oblast's lead in the mid-1990s in attempting to bargain for greater economic autonomy, was always more discreet with the Kremlin. Governor Petr Sumin expressed full support for President Putin's federal reforms and has maintained respectable relations with Presidential Envoy Latyshev from the very beginning of the latter's appointment.

During 2000-03, the Urals region, along with many other Russian regions, found itself in the midst of a "third wave" of property redistribution; however, this wave showed a much greater involvement of national oligarchs and FIGs. Chelyabinsk Oblast, until recently, was not on the radar screen of the national FIGs. This situation changed in late 2000, when the oblast's metallurgical, machine-building and chemical industries began to register the first signs of growth in output and profitability and became attractive to Moscow FIGs,[10] especially selected plants in Magnitogorsk, Miass, and Zlatoust.[11]

Conflicts involving the metal-producing and machine-building industries, as well as the presence of influential national FIGs have been familiar sights in Sverdlovsk Oblast even before the "third wave."[12] In 1999, over 40 percent of industrial output in the oblast was controlled by transregional and national FIGs.[13] The involvement of regional and national oligarchs in regional politics was not the news either. It was reported that the well-known oligarch, Kakha Bendukidze, owner of the machine-building giant United Machine-Building Plants Holding Company (Obyedinennye mashinostroitelnyye zavody), which includes UralMash,

---

[8] ITAR-TASS, November 24, 1998.

[9] Interview with V. Kulikov, Deputy Director, Ministry of Foreign Affairs Department on relations with NGOs, political parties, and Russian regions, November 2000.

[10] In April 2001, for example, Siberian Aluminum (SibAl), the Chelyabinsk oblast administration, and the Urals Automobile Plant (UralAZ) signed an agreement concerning SibAl investment into UralAZ. SibAl guarantees certain level of wages and social benefits for the plant's workers, contributions to social infrastructure and the Miass city budget. The Chelyabinsk oblast government will monitor SebAl activities in the oblast. In addition, investments would be channeled through Chelyabinsk oblast banks. ("Sumin budet kontrolirovat Deripasku," Strana.ru website, April 17, 2001, <http://www.strana.ru>.)

[11] Marat Chaplygin, "Mikhail Grishankov: Zashchitim Chelyabinskuyu oblast ot nashestviya oligarkhov," *Argumenty i Fakty* (Chelyabinsk edition), No. 31, August 2000, p. 8.

[12] Konstantin Simonov, "Na rodine Yeltsina gotovyatsya k gubernatorskim vyboram," *Russkaya Mysl*, June 10–16, 1999, <http://www.ancentr.ru>.

[13] Dmitriy Tolmachev, "Chem menshe igrokov, tem legche dogovoritsya," *Ekspert-Ural*, March 12, 2001; in Integrum Techno, <http://www.integrum.ru>.

Izhorskiye Zavody, and Krasnoye Sormovo, provided "significant support" to Eduard Rossel during his 1999 election campaign.[14]

The fight over lucrative businesses in the Urals among local, regional, and national oligarchs (in some cases even foreign companies) intensified after 2000. Several conflicts, such as the one over the metallurgical plant in Nizhnyaya Salda and the Sverdlovsk Machine-Building Plant, involved the use of firearms in disputes over ownership.[15]

In light of the federal government's plans to form several large corporations in the defense industry and sell off unprofitable defense enterprises, confrontation between regional and national oligarchs in the Urals is likely to become even more acute in the coming years. Fortunately, none of the Minatom nuclear weapons complex facilities is currently on the list for privatization.[16]

The Minatom Complex in the Urals

To fully understand the processes that have unfolded around Minatom, nuclear policy, and nuclear facilities in the region, one must unravel a complex network of horizontal and vertical ties among nuclear facilities, closed city administrations, regional authorities, other regional actors, and the federal government as a whole, including the presidential representative to the federal district. This section will examine the existing legal framework in the nuclear sphere in Chelyabinsk and Sverdlovsk Oblasts, key areas of interaction with the regions, as well as other local, regional, and federal variables that affect nuclear safety and security.

Legal Framework

During the Yeltsin era, Sverdlovsk and Chelyabinsk Oblasts both concluded general power-sharing agreements with the federal government in 1996 and 1997, respectively, which were followed by a number of supplemental agreements in specific areas. During the course of Putin's federal reforms, these power-sharing agreements between Moscow and the regions were allowed to expire or encouraged to be abrogated. Chelyabinsk Oblast Governor Sumin announced its oblast agreement void in early 2001 in response to Moscow's request, while Sverdlovsk Oblast Governor Rossel resisted the nullification of the agreement until February 2002 and agreed to its cancellation only under direct pressure from

---

[14] Ivan Yudintsev, "Po tu storonu sormovskikh barrikad," *Monitor (N.Novgorod)*, March 20, 2000.

[15] Aleksandr Polozov, "ALSTOM i Rossel dogovorilis o sudbe CHEMZa," Strana.ru website, July 5, 2001, <http://ural.strana.ru>, and Maksim Pokrovskiy, "Zachem deputatu zavod," Strana.ru website, July 5, 2001, <http://ural.strana.ru>.

[16] Natalya Kulakova and Yelena Kiseleva, "Prodayetsya kusok gosudarstva," *Kommersant-daily*, August 16, 2001.

Moscow.[17] Nevertheless, it is worth examining major provisions of these agreements, since they laid the foundation for regional legislation in the nuclear area and shaped federal legal provisions concerning use of nuclear energy and the import of spent nuclear fuel.

The 1996 agreement between Sverdlovsk Oblast and the federal government on the defense industry stipulated that the authority to regulate and coordinate civilian production and conversion at defense facilities was to be shared between federal and regional offices. The regional government had a right to "take part in making decisions regarding facility closures" and to provide funds for defense enterprises if federal funds were unavailable. In the latter case, regional offices were entitled to reduce tax payments to the federal budget.[18] According to Article 3 of the agreement, directors of state-owned defense enterprises in the oblast were to be appointed with the consent of the Sverdlovsk Oblast government. In addition to the power-sharing agreements, the two oblasts passed a number of their own laws on nuclear issues. In general, these laws adhered to the regulations stipulated by the power-sharing agreements and federal legislation in the nuclear sphere.

Several specific norms from the power-sharing agreements and the oblast legislation in the nuclear sphere deserve a more detailed review. One such norm was contained in the power-sharing agreement on nuclear energy between Chelyabinsk Oblast and the federal government. It required a mandatory allocation of 25 percent of all proceeds from the reprocessing of imported spent nuclear fuel at Mayak for environmental and social programs in Chelyabinsk Oblast so that 12.5 percent of proceeds would remain at Mayak and 12.5 percent would be allocated to a special hard-currency fund managed by the oblast government for social programs and environmental rehabilitation.[19] Another example from the same agreement is an allocation of 10 percent of construction costs for any new nuclear facility or activity in the Chelyabinsk Oblast for the development of the adjacent territory.[20]

In addition to power-sharing agreements, regional governments concluded individual agreements with Minatom, its nuclear enterprises, and, in some cases, closed cities. These agreements define the status of enterprises, long-term financial and social commitments of both sides, or deal with certain short-term problems, such as restructuring nuclear facilities' debts or directing financial contributions from each side into specific projects. An agreement between Minatom and the

---

[17] Kremlin denied Rossel meetings with Putin until he denounced the agreement. See Channel 4 TV news report from March 7, 2002, <http://www.channel4.ru>.

[18] Article 1 of the Agreement between the Government of the Russian Federation and the Government of Sverdlovsk oblast on sharing powers in the defense industry sector, December 1, 1996; in Integrum Techno, <http://www.integrum.ru>.

[19] Article 5 of Agreement No. 19 between the Government of the Russian Federation and the Chelyabinsk Oblast Administration on sharing powers in the sphere of the use of atomic energy at federal and interregional facilities, June 13, 1998.

[20] Article 3 of Agreement No. 19 between the Government of the Russian Federation and Chelyabinsk Oblast Administration on sharing powers in the sphere of the use of atomic energy at federal and interregional facilities, June 13, 1998.

Sverdlovsk Oblast for the construction of the BN-800 unit at the Beloyarsk Nuclear Power Plant in Zarechnyy specified mutual obligations for Minatom and the oblast regarding funding, subcontractors, and the construction schedule.[21] Another example was an agreement between Sverdlovsk Oblast, Minatom, and the Urals Electrochemical Combine in Novouralsk that spelled out procedures for establishing joint funding for the construction of a cancer center in Yekaterinburg.[22]

Key Areas of Strategic Interaction

The most active and conspicuous interaction between the nuclear complex facilities, closed cities, and the regional authorities in the Urals took place on economic, social, and safety and security issues.

Economic and Financial Relations

The direct financial contributions made by nuclear facilities to oblast budgets have been minimal. According to federal legislation, the vast majority of taxes collected in closed cities (Russian acronym—ZATO) are remitted either to the federal budget or local coffers, with only a small share allocated to regional budgets.[23] Indirect contributions, on the other hand, have been much more significant.

To begin with, the closed cities' budgets are on a scale comparable to entire oblast budgets. For example, in 1999, the combined budget of the four closed cities in Chelyabinsk Oblast (i.e., Ozersk, Snezhinsk, Trekhgornyy, and Lokomotivnyy) was roughly equivalent to the entire oblast budget (9 and 11.5 billion rubles, respectively).[24] Although this imbalance created tension between regional authorities and the closed nuclear cities, it benefited the oblasts in several ways—creating islands of social stability, providing investment for developing regional social infrastructure, and providing employment to neighboring territories. This was especially true for new construction, such as the BN-800 unit at the Beloyarsk Nuclear Power Plant or the construction of the Fissile Material Storage Facility at Mayak. Another illustration of economic spillover to adjacent territories took place in Trekhgornyy, which employed approximately 2,000 workers from nearby villages and towns. In 1999 and 2000, the city of Trekhgornyy spent 14 million

---

[21] Vladimir Bystrykh, "Chetvertyy energoblok Beloyarskoy AES budet stroit trest 'Uralenergostroy,'" Strana.ru website, March 6, 2001, <http://www.strana.ru>.

[22] Interview with Eduard Rossel, "Luchshe druzhit domami," *Vek* online version, July 13, 2001, <http://www.wek.ru>.

[23] The only exception of direct contribution to oblast budget is the described above mandatory allocation of 12.5 percent of proceeds from foreign spent nuclear fuel reprocessing.

[24] Igor Stepanov, "Kak podelit dengi zapretok?" *Delovoy Ural*, April 27, 2000.

and 25 million rubles, respectively, for social and economic programs in adjacent territories.[25]

Tension between regional authorities and closed cities on budgetary and tax issues was particularly acute during the "offshore period" within the ZATOs. From 1997 to 1999, the closed cities were allowed to provide tax benefits to, and keep taxes collected from, enterprises registered within the city limits.[26] There were numerous violations resulting in significant tax revenue losses at all levels, as many regional and national businesses seized this opportunity to lower their individual tax liability. However, the lack of clear guidelines and precisely defined legal procedures ensured that on many occasions taxes were not collected, diverted to wasteful projects, or ended up in the pockets of local bureaucrats. Chelyabinsk Oblast Governor Sumin was very critical of the offshore zones in Chelyabinsk Oblast's closed cities, especially when many major contributors to the oblast budget re-registered there. He threatened "to use every available resource" to make sure that oblast enterprises no longer hid their taxes in closed cities, and most of the oblast businesses moved their registrations back.[27]

Despite the criticism, federal support was critical for allowing most of the cities to improve their social programs and infrastructure and to assist with the operation of nuclear facilities in their territories. Several closed cities in the Urals used the extra money as originally intended (i.e., for business development and economic diversification). The most successful example is Trekhgornyy in Chelyabinsk Oblast where the Mayor Nikolay Lubenets was instrumental in developing new businesses, including conversion projects at the Instrument-Making Plant.[28] Although some of these projects were controversial and Lubenets was harshly criticized for abusing the city's offshore status, Trekhgornyy was able to achieve financial independence from Moscow and has not received subsidies from the federal budget since 2001.

The only other closed city in the Urals that has not received federal subsidies is Novouralsk in Sverdlovsk Oblast.[29] Novouralsk was able to achieve this distinction with the help of proceeds from the U.S.-Russian HEU-LEU deal and other export-oriented uranium enrichment services. By comparison, the Instrument-Making Plant in Trekhgornyy, which is involved in the assembly and dismantlement of nuclear warheads, does not have such lucrative contracts.

---

[25] Interview with Trekhgornyy Mayor Nikolay Lubenets published in *Yuzhnouralskaya panorama* online edition, December 9, 2000, <http://www.chelpress.ru/newspapers/panorama/>.

[26] Snezhinsk and Sarov retained this special tax regime in 2000 as well.

[27] Anatoliy Khodorovskiy, "Legche stanet cherez 2–3 goda," *Vedomosti*, August 17, 2000, p. A5.

[28] Olga Sokolova, Igor Stepanov, "Sekret uspekha 'zakrytoy' ekonomiki," *Delovoy Ural*, December 21, 2000.

[29] Vladimir Dernovoy, "Zagadka 'sekretnykh' gorodov. ZATO sposobny sozdavat unikalnyye tekhnologii," *Vek*, August 28, 2001. Federal Budget of the Russian Federation for the year 2001, *Sobraniye zakonodatelstva RF*, No. 1, January 1, 2001.

Another positive aspect associated with the offshore period in 1997-99 was the financial assistance that nuclear facilities received from city administrations via credits to pay wages, deferrals of payments to the local budget (or even write-offs of long overdue payments), and tax exemptions.[30] In August 1998, the Snezhinsk City Council approved a 7.5 million ruble credit for VNIITF.[31] In November 1998, the mayor of Snezhinsk wrote off VNIITF's debt to the city budget that was accumulated between 1994 and 1997.[32] Summing up the relations between the city administration and the Instrument-Making Plant during that period, Trekhgornyy Mayor Lubenets stated that "in actuality, strategic weapons manufacturing was paid for by the local budget via these subsidies to the nuclear facilities."[33]

Without a doubt, relations between city administrations and nuclear facilities are very important for both sides in the shaping of the social and economic environment in the closed cities, especially since the majority of city residents are either employed by the nuclear facility or are relatives of those who work there. Creation of civilian businesses is one of the areas where joint facility and city efforts seem to accelerate conversion and minimize social and economic stresses from downsizing, whereas disagreements between city and facility administrations became obstacles to retooling closed cities and downsizing the Russian nuclear complex.[34] Sources, in both in the United States and Russia, noted that one of the impediments to the progress of the U.S.-funded Nuclear Cities Initiative and its goal of economic diversification in Snezhinsk has been the tense relationship between VNIITF Director Georgiy Rykovanov and the head of the city administration Anatoliy Oplanchuk, as well as "quiet resistance" from the VNIITF administration to conversion and downsizing projects.[35]

The Chelyabinsk Oblast administration, in an effort to compensate for the loss of money to the ZATO offshore zones, established a special extra-budgetary fund—the "Fund for the Development of the Territories"—to which all closed cities were supposed to contribute. The money was intended to upgrade the region's infrastructure and to support road improvements in territories neighboring the closed cities. The deal was negotiated with all three Minatom closed cities in Chelyabinsk Oblast. However, because the oblast administration managed the fund poorly, the amount that was contributed to it by the closed cities was much lower than expected. The Chelyabinsk Oblast legislature made several unsuccessful

---

[30] This is particularly true for Trekhgornyy and Snezhinsk in Chelyabinsk oblast.

[31] "Postanovleniya Gorodskogo Soveta Narodnykh Deputatov g. Snezhinsk," City Council publications, Snezhinsk, 2000.

[32] "Vsem, komu dolzhny, prostili," *Okno*, November 26, 1998, p. 2.

[33] Yuliya Latynina, "Ray za kolyuchkoy," *Ekspert*, No. 20, June 1, 1998.

[34] Positive results in economic diversification and conversion in Trekhgornyy, Chelyabinsk Oblast and Zarechnyy, Sverdlovsk oblast are attributed, to a certain degree, to very active city mayors and administrations. However, without consensus with nuclear facilities' management, it is hardly likely these results could have been achieved.

[35] This opinion was shared on condition of anonymity by a U.S. representative involved in business development projects in Russia's closed cities and by several businessmen from Snezhinsk.

attempts to approach nuclear facilities to collect the promised contributions, but, by that time, tax privileges to ZATOs had been eliminated by the federal government and the fund gradually ceased to exist.[36]

Mandatory payments to the oblast budget from the proceeds for foreign spent fuel reprocessing at Mayak vanished by 2000 due to the changes in the Russian environmental legislation and the subsequent loss of foreign spent fuel contracts. In response, the Chelyabinsk Oblast legislature petitioned the Russian government demanding mandatory payments from the reprocessing of domestic spent nuclear fuel at Mayak, including spent naval fuel.[37] This request, however, was never granted.

From the perspective of the oblast, ZATOs were not always considered "cash cows." In the early and mid-1990s, when nuclear enterprises and closed cities were enduring grave socio-economic conditions, regional authorities stepped into the fray by providing loans to pay employees and by calling to the attention of the federal government the hardship facing nuclear facilities and their employees. Sverdlovsk Oblast, for example, provided subsidies to the closed cities of Novouralsk and Lesnoy in 1995-96.[38]

In February 1998, the Chelyabinsk Oblast legislature issued an appeal to the Federal Assembly of the Russian Federation regarding the critical situation in the nuclear defense complex, including three Chelyabinsk Oblast nuclear enterprises.[39] In April 1999, during a meeting with President Yeltsin, Chelyabinsk Oblast Governor Sumin complained that the federal government had failed to pay to its nuclear enterprises for defense orders. He pointed out that Chelyabinsk Oblast simply did not have the resources to support these enterprises, nor was it supposed to support enterprises owned by the federal government.[40]

Despite the relatively clear delineation of authority and responsibilities between the center and the two oblasts and the stabilization of federal payments to nuclear facilities in 2000-01, nuclear facilities continue to view oblast governments as supporters in voicing their facilities' concerns at the national level. For example, in April 2001, Governor Rossel sent a letter to the Russian prime minister on behalf of the Urals Electromechanical Plant in Yekaterinburg requesting that the facility's

---

[36] Ibid; and "Vsyudu dengi, gospoda," Ozersk.ru website, July 14, 2000, <http://www.ozersk.ru>.

[37] "K prezidentu Rossiyskoy Federatsii o deyatelnosti proizvodstvennogo obyedineniya 'Mayak' i radiatsionnoy bezopasnosti Uralskogo regiona," Decree No. 845, Legislative Assembly of Chelyabinsk Oblast, April 27, 2000.

[38] Sverdlovsk Oblast Law No. 57-O3 from December 24, 1996 on the oblast budget for 1997, *Sobraniye zakonodatelstva Sverdlovskoy oblasti*, No. 8, 1997, p. 1163.

[39] "Obrashcheniye Zakonodatelnogo sobraniya Chelyabinsky oblasti 'k Federalnomu Sobraniyu Rossiyskoy Federatsii o polozhenii v yadernom oruzheynom komplekse Rossii'," February 26, 1998; in *Yadernyy Kontrol* No. 3 (May-June 1998), p. 31.

[40] Igor Stepanov, "Chelyabinsk Governor Focuses on Nuclear Weapons in Yeltsin Meeting," *Russian Regional Report*, 4, April 21, 1999. Regional experts mention that in 1997 the Chelyabinsk Oblast government offered to financially support the closed nuclear cities in exchange for the right to collect and keep taxes in closed cities.

federal tax debt be restructured, given that it was accumulated in 1994-99 mostly due to the failure of the federal government to pay for defense orders.[41] Another example was the complaint by the Novouralsk city administration to Rossel about the loss of part of revenues from the HEU-LEU deal and export contracts because of a new mechanism of the distribution of proceeds between enrichment facilities and Minatom introduced in 2001. This new system significantly decreased payments to the Urals Electrochemical Combine, which in turn cost the Novouralsk city budget 200 million rubles in 2001.[42] These appeals to regional authorities suggested the limits to re-centralization efforts. Relations between facilities and regions have probably reached a point of no return in the sense that local facilities will continue to enlist the help of regional governments in solving their financial problems when Minatom or the federal government fail to meet their obligations.

Non-Budgetary Relations

Nuclear facilities and closed cities also affect regional economies via contributions to infrastructure, employment, and social programs. As mentioned above, the Chelyabinsk Oblast administration used various regional funds to cover development and renewal programs for the territories adjacent to closed cities. In Sverdlovsk Oblast, the approach was somewhat similar, as the regional government negotiated a deal with Minatom and Novouralsk to fund the construction of a cancer center in Yekaterinburg, which opened in 2000. Minatom also sponsored the construction of a sports center in Yekaterinburg and other social projects.[43]

Both oblasts have experienced severe energy shortages, as indigenous suppliers have been unable to keep up with the heavy loads demanded by respective industrial complexes. Chelyabinsk Oblast's energy resources, for example, support roughly 60 percent of demand.[44] Moreover, the future economic prosperity of the two oblasts depends on cheap and plentiful energy resources to ensure the competitiveness and profitability of their industries. Accordingly, both governments have been strong advocates of Minatom's plans to expand nuclear power production and development in their territories,[45] expecting to benefit not only from stable supply and lower energy prices, but also from employment and

---

41 Biznes-Novosti Urala, No. 144, April 9, 2001; in Integrum Techno, <http://www.integrum.ru>.

42 "Gubernator Sverdlovskoy oblasti posetil predpriyatiya Novouralska," Biznes-Novosti Urala, January 15, 2001; in Integrum Techno, <http://www.integrum.ru>.

43 One more medical center, financed by Minatom, is under construction. Viktor Vepritskiy, "Luchshe druzhit domami," *Vek* online version, July 13, 2001, <http://www.wek.ru>.

44 "Generalnyy director PO 'Mayak' Vitaliy Sadovnikov schitayet neobkhodimym vozobnovit stroitelstvo Yuzhno-Uralskoy AES," Neftegazovaya vertical, June 26, 2001, <http://www.ngv.ru>.

45 Not all members of the oblast legislatures support nuclear power plant construction or other Minatom projects, but they are a small minority and cannot alter the final decision.

business opportunities associated with the construction of nuclear power plants, including contracts for local construction, machine-building and metal-processing enterprises.

In 2001, Minatom announced its decision to start the construction of a new unit, the BN-800 fast breeder reactor, at the Beloyarsk Nuclear Power Plant located in Sverdlovsk Oblast. This decision was received with great enthusiasm by the regional government.[46] The Chelyabinsk Oblast government was less "lucky" and continues to plead with the federal government and Minatom to resume construction of the South Urals Nuclear Power Plant (NPP) in Ozersk, where work was halted after pressure from local environmental groups and the public following the Chernobyl accident. Officials at Mayak and the Chelyabinsk Oblast have argued that the nuclear power plant would provide employment to Mayak specialists, solve the oblast's energy problem, and help evaporate radioactive water from the artificial reservoirs on the territory of Mayak.

Safety and Security

Another area in which oblast governments share powers and responsibilities with various federal government agencies is with respect to the safe operation and physical protection of nuclear facilities. Oblast governments have focused primarily on undertaking measures to facilitate accident prevention, the readiness of emergency response units, and physical protection of nuclear facilities from terrorist attacks. The involvement of oblast governments in activities to address these issues from inside the nuclear facilities is rather limited, as Minatom and other federal agencies are in charge of these issues. However, the role of oblast and local governments should not be underestimated, especially in oblast or region-wide systems of emergency response, security, law enforcement, export control, and customs.

Governors and oblast governments often initiate and coordinate training exercises and provide funding for emergency response systems. Attention to emergency response units and antiterrorism exercises intensified in both oblasts in the wake of terrorist attacks in Russia in 1999 and 2000 and with the renewed war in Chechnya.[47] Similarly, there was a surge of meetings and training exercises at the oblast and regional levels following the September 11, 2001 attacks in the United States. Emergency preparedness and physical protection also are regular items on the agenda of oblast government meetings.[48]

---

[46] Viktor Vepritskiy, "Luchshe druzhit domami," *Vek* online version, July 13, 2001, <http://www.wek.ru>.

[47] For examples, see "Eduard Rossel podpisal ukaz 'O merakh po obespecheniyu antiterroristicheskoy i protivodiversionnoy zashity obektov i bezopasnosti grazhdan v Sverdlovskoy oblasti,'" Region-Inform, February 29, 2000, <http://www.rossel.ru>.

[48] For example, see "Pravitelstvo Sverdlovskoy oblasti utverdilo plan deystviy po zashchite naseleniya, prozhivayushchego ryadom s BAES," Region-Inform, November 9, 1999. See also the report on the review of antiterrorism training of nongovernmental security

Nuclear cities and facilities often address their specific concerns to both oblast and federal authorities. In March 2001, residents of Zarechnyy—after the announcement of the beginning of the construction at the Beloyarsk NPP—sent a petition to Governor Rossel and to Presidential Representative in the Urals District Latyshev requesting tightened security in the city because of the expected inflow of outside workers during the construction of the new unit.[49] Moreover, in the late 1990s, the Sverdlovsk Oblast government assisted the Elektrokhimpribor Combine in launching conversion projects, receiving credits, and, most importantly, pressuring Minatom to finish the construction of a new storage facility for nuclear warheads and their components at Elektrokhimpribor.[50] The new warhead storage facility was intended to replace the old storage site, "which did not meet the requirements of international organizations."[51]

Among the most serious concerns related to the safety and security of nuclear materials and weapons is the reliability of the military and civilian personnel that guard these nuclear facilities and closed cities. Practically all closed cities in the Urals have their own lists of thefts, shootings, or other accidents in which guards from military units or hired personnel were responsible. Nuclear facilities and, in some instances, closed city administrations supplement the income of the military and improve their housing and living conditions, in part to prevent such incidents. In March 2002, the governor of Chelyabinsk Oblast announced that the oblast government would provide financial and technical support to the units of the MVD in the Urals that protect nuclear facilities in Sverdlovsk and Chelyabinsk Oblasts.[52]

The effectiveness of the export control system is another important issue that concerns regions with nuclear materials and technologies, especially in those areas where cases of illegal operations or thefts of nuclear material have been registered. The Urals is one of these regions. In 1997, illegal export operations involving iridium-192 were uncovered at Mayak. The operations came to light only when a regular shipment from Mayak to the United Kingdom was filed at the St. Petersburg customs office instead of the Kyshtym customs office in Chelyabinsk Oblast.[53] In 2000, one of the first actions of the presidential representative in the Urals District was the replacement of several senior regional customs and law

---

personnel by the Chelyabinsk Security Council committee, Alla Skripova, "Antiterroru budut uchit," *Vecherniy Chelyabinsk*, June 7, 2001.

[49] "Zhiteli Zarechnogo trebuyut usilit mery po okhrane pravoporyadka v gorode," Strana.ru website, March 20, 2001, <http://www.strana.ru>.

[50] Stanislav Lavrov, "Kogda otkazyvayut tormoza," *Oblastnaya gazeta*, March 14, 2000; in Integrum Techno, <http://www.integrum.ru>; and "Gubernator E. Rossel i Ministr po atomnoy energii Ye. Adamov 21 maya pobyvali na kombinate 'Elektrokhimpribor' v Lesnom," Sverdlovsk Governor's Office press release, May 21, 1999, <http://www.rossel.ru>.

[51] Ibid.

[52] "Chelyabinskaya oblast. Gubernator Petr Sumin vkladyvayet dengi v bezopasnost yadernykh obektov," Regions.Ru website, March 14, 2002, <http://www.regions.ru>.

[53] Viktor Riskin, "Mechenyye Izotopy," *Chelyabinskiy rabochiy*, June 16, 1997, p. 2.

enforcement officials who were considered "too closely associated" with governors or other oblast structures.

Energy shortages are also closely linked to the safety and security of nuclear facilities in the region. The power outage in September 2000 in the South Urals provides a serious warning about the risks posed by energy problems. A malfunction in the Southern Ural Power Grid led to a power outage at Beloyarsk NPP (Zarechnyy) and Mayak (Ozersk). Both facilities had to shut down operating nuclear reactors manually in order to avoid major accidents. According to some reports, manual interference became necessary when it turned out that the backup diesel generators were not operational.[54] Mayak faced a similar situation during the 1998 strike of Chelyabinsk coal miners, who blocked the railway leading to the Argayash Power Plant that provides power and heat to the complex.[55]

Energy problems in the South Urals also are aggravated by debts owed to regional power providers. This was particularly acute in the late 1990s and early 2000s. There have been no reports to date about intentional cutoffs of electricity to nuclear facilities in Chelyabinsk and Sverdlovsk Oblasts (federal legislation prohibits power cutoffs to strategic facilities) due to non-payments. Nevertheless, in May 2001, Ural-Press-Inform reported that Chelyabenergo (the regional power provider) cut back the energy supply to Instrument-Building Plant in Trekhgornyy and VNIITF in Snezhinsk by 30 percent because of systematic non-payment.[56] In September 2001, on the eve of a visit by Minister of Atomic Energy Alexander Rumyantsev to Chelyabinsk Oblast, Chelyabenergo cut the energy supply to Mayak by 30 percent due to its electricity debts.[57]

Nuclear facilities depend on a steady supply of power both for safe operation and for the physical protection of nuclear weapons and fissile materials. Although the situation with energy supplies is not as severe as in the Russian Far East, it is far from secure. Chelyabinsk Oblast imports up to 40 percent of its power. In addition, the equipment and infrastructure of the regional energy company Chelyabenergo are outdated and require replacement, as one-fifth of this

---

[54] Sergey Blinovskikh, "Ostanovleny dva reaktora—chrezvychaynaya situatsiya v PO 'Mayak'," *Chelyabinskiy rabochiy*, September 12, 2000; in Integrum Techno, <http://www.integrum.ru>. "Ural Regions Blackout Stops Nuclear, Metallurgical Plants," ITAR-TASS, September 11, 2000; in RANSAC Nuclear News, September 11, 2000. Amelia Gentleman, "Nuclear Disaster Averted—Russian Power Plant Workers Praised for 'Heroic' Operation to Cool Reactors," *Observer*, September 17, 2000; in RANSAC Nuclear News, September 18, 2000. Yelena Vyatkina, "35 minut v 'otklyuche', ili 'Mayak' v napryazhenii i bez nego," PO Mayak website, September 18, 2000, <www.ozersk.ru/mayak/news/news.shtml>.

[55] *Vesti*, August 5, 1998; in "Russian Nuclear Facility Affected by Miners' Rail Blockade," FBIS-SOV-98-217.

[56] "V Chelyabinskoy oblasti energetiki otkluchili 849 potrebiteley elektroenergii," Ural-Press-Inform, May 25, 2001; in Integrum Techno, <http://www.integrum.ru>.

[57] "OAO 'Chelyabenergo' ogranichilo za dolgi energosnabzheniye yadernykh tsentrov Chelyabinskoy oblasti," Rosbizneskonsulting, September 25, 2001; in Minatom news digest, September 27, 2001, <http://www.minatom.ru/>.

equipment has already exceeded its planned service life.[58] A similar, though less acute, situation persists in Sverdlovsk Oblast.

Environmental Issues

Economic, safety, and security issues in the Urals are tightly intertwined with environmental problems and concerns. The Urals region has an infamous history of accidents and radioactive contamination, especially in Chelyabinsk Oblast. The most prominent nuclear accident occurred in 1957 at the Mayak Facility. Another notorious example of the environmental legacy of nuclear programs at Mayak is Lake Karachay, which was used by the complex for dumping highly radioactive waste and which continues to pose a threat to the environment. In addition, Mayak continues to discharge liquid radioactive waste into its surface reservoir system which has already accumulated over 400,000 cubic meters of liquid radioactive waste.[59]

The nuclear legacy in the region and environmental risks associated with the operation of existing nuclear facilities has remained on the radar screens of regional and local governments, the general public, and regional environmental organizations. The environmental movement in Chelyabinsk Oblast is one of the most active anti-nuclear movements in Russia. At least six organizations in Chelyabinsk Oblast focus on nuclear safety, security, and environment, with the Movement for Nuclear Security leading their efforts. Environmental organizations employ a variety of methods to attract attention to nuclear problems in the region—from filing legal claims against Mayak to picketing government buildings to organizing anti-nuclear camps. Most of the anti-nuclear and environmental activities in Chelyabinsk Oblast are focused on Mayak. In turn, the Mayak administration blames environmental organizations for the plant's financial problems, allegedly because of the halt in the construction of the South Urals NPP and the loss of revenues associated with the processing of foreign spent nuclear fuel. The administration goes even further and claims that regional environmentalists are one of the major threats to nuclear safety and security because of their extreme protest campaigns.

The Chelyabinsk Oblast government has no warm feelings for the environmental organizations either. Statements by oblast FSB officials labeling these groups' activities as "gathering classified information" and "harming the oblast's image" only reinforce the oblast's negative attitude toward environmental

---

[58] Igor Stepanov and Vladimir Terletskii, "Gonki po vertikali," *Ekspert-Ural*, April 23, 2001.

[59] Interfax, May 12, 1996; Vladislav Larin, "Three Radiation Disasters At Mayak Integrated Plant," *Energiya: Ekonomika, Tekhnika, Ekologiya*, March 1, 1996, pp. 46-53; in "Documents Reveal Details Of Urals Nuclear Disaster," FBIS-UST-97-002, January 14, 1997.

groups.[60] At the same time oblast governments use environmental arguments to lobby Moscow. The Chelyabinsk government sent numerous petitions to the federal government about the threat posed by contaminated water in Mayak's system of artificial water reservoirs also known as the Techa cascade of lakes. The lakes are saturated with liquid radioactive waste and could leak into the Iset, Tobol, and Ob Rivers and flow into the Arctic Ocean. The oblast has suggested four options to address this problem. Its preferred option, though, is resuming construction of the nuclear power plant in Ozersk, which supposedly would use the water from the reservoirs to cool the reactor.[61]

The environmental movement in Sverdlovsk Oblast is much weaker and less organized than in Chelyabinsk Oblast. Cooperation between the two oblasts' groups also is rather weak. To a certain degree, the more tempered debate over nuclear and environmental issues in Sverdlovsk Oblast can be attributed to more effective public relations by regional officials. During the 1999 gubernatorial election campaign, Rossel's opponents claimed that he issued a permit to build a nuclear waste storage facility in Lesnoy. Immediately, several articles appeared in the regional press that rebuffed such allegations and characterized the Lesnoy project as a modern storage facility for nuclear warheads and components that actually improves the state of nuclear safety and security for the oblast.

Despite pro-environment statements by both oblast governments, the environment is often sacrificed for short-term economic benefits or used as a bargaining tool vis-à-vis the federal government and international assistance providers. This was particularly evident during the debates over the import of spent nuclear fuel. Mayak is the only Russian facility currently involved in the reprocessing of spent nuclear fuel from power and naval reactors. As mentioned above, a special agreement (effective until 2001) between Chelyabinsk Oblast, Minatom, and Mayak stipulated the allocation of 12.5 percent of proceeds from reprocessing to the oblast budget for environmental and social programs. The 1991 Russian law on environmental protection prohibited the import and storage of foreign nuclear waste in Russia, and banned contracts for reprocessing of foreign spent nuclear fuel from the East European countries which still operate Soviet-built nuclear power reactors. Consequently, Mayak gradually lost its contracts for the reprocessing of foreign spent fuel, which had earned the facility roughly $50-60 million in annual revenues.[62] Nearly 98 percent of Chelyabinsk Oblast's hard currency revenues in the mid-1990s came from spent fuel reprocessing proceeds at

[60] For example, see the letter from former Head of FSB in Chelyabinsk oblast V. Tretyakov to Governor Sumin "Razmyshleniya kontrrazvedchika 'Ob ekologicheskikh faktorakh bezopasnosti regiona'," *Zelenyy mir*, No. 26 (346), 2000.

[61] The authorship of this idea belongs to Mayak.

[62] Marina Zaytseva, "Khimkombinat 'Mayak' teryayet zarubezhnyye kontrakty," Regions.ru website, March 15, 2000, <http://www.regions.ru>; Igor Stepanov, "Dorogoye 'udovolstviye'," *Magnitogorskiy rabochiy*, on-line version, May 17, 2000, <http://gazeta.magnitka.ru>.

Mayak.[63] This explains to a large degree why Chelyabinsk legislators and representatives from the oblast government supported Mayak's and Minatom's efforts to amend Russian environmental laws to allow import and reprocessing of foreign spent nuclear fuel, despite the controversy surrounding this issue and the unpopularity of this idea among the population. The shifts in the level of support for spent fuel imports among oblast officials followed the debates over the size of financial compensation to region for the risks involved. Deputy Governor of Chelyabinsk Oblast Gennadiy Podtyesov warned early on that if the oblast did not receive 30 percent of the spent fuel proceeds, the administration would ban shipments of foreign spent fuel into the region.[64] Accordingly, the new legislation stipulated that 25 percent of the proceeds (after reimbursement for expenses associated with the storage and reprocessing) should be allocated to budgets of regions housing reprocessing facilities.[65]

In Sverdlovsk Oblast, which does not have spent fuel reprocessing facilities, the reaction of the oblast legislature to the Minatom-sponsored amendment turned from neutral at the early stages to negative when it became clear that most likely the only beneficiaries of the spent fuel deal would be those oblasts directly involved in reprocessing. In February 2001, the Sverdlovsk Oblast legislature sent a petition to the Federal Assembly and Russian president questioning the legislative amendment allowing import of spent fuel.[66]

As Russian regions become more knowledgeable about legal procedures, gaps in current legislation, especially in areas of joint or not clearly defined jurisdiction in the nuclear sphere, have the potential to become the next hot issue of debate between Moscow and the regions. According to current Russian legislation, construction of any new nuclear facility must be approved by the regions and municipalities affected.[67] Although current Russian legislation does not specify the procedures for such approval or the enforcement mechanisms, the required consent from the region has the potential to become a powerful lever that regions can use on Minatom and the federal government.

Besides the previously mentioned economic aspects of regional responses to the law on the import of spent nuclear fuel, one more illustration highlights the existing shortcomings of nuclear legislation in Russia with respect to the interests of the regions: the process of planning and negotiating nonproliferation assistance programs. This can be illustrated by an attempt by the Chelyabinsk Oblast administration to demand a percent of the construction costs of the Fissile Material

---

[63] Ibid.

[64] Aleksandr Polozov, "V Chelyabinske rasschityvayut na shchedroye voznagrazhdeniye za utilizatsiyu inostrannykh yadernykh otkhodov," Strana.ru website, February 22, 2001, <http://www.strana.ru>.

[65] See Articles 4 and 5 (item 5) of the Law of the Russian Federation N 92-FZ on Special Environmental Remediation Programs for Land Contaminated with Radiation, July 10, 2001.

[66] "Obrashcheniye deputatov Sverdlovskoy oblastnoy dumy k rukovodstvu Rossiyskoy Federatsii," February 22, 2001, Strana.ru website, <http://www.strana.ru>.

[67] See Article 28 of the Law on Atomic Energy, N 170-FZ, November 21, 1995.

Storage Facility (FMSF) at Mayak, funded by the United States under the Cooperative Threat Reduction Program. The facility is designed to store plutonium from dismantled nuclear warheads. The agreement to build the storage facility was negotiated by the U.S. Department of Defense with Minatom in 1992. After failing to secure an agreement from Tomsk Oblast to host such a facility, Minatom turned to Chelyabinsk Oblast and promised to fund social programs and infrastructure upgrades in territories neighboring Mayak in return for its hosting of the facility. The 1998 power-sharing agreement in the nuclear area between the Russian government and Chelyabinsk Oblast required the allocation to the oblast of 10 percent of construction costs from any new nuclear facility. In 2001, when the FMSF was almost completed, the oblast requested the Russian government to remit to the oblast budget 10 percent of the facility's construction costs, which at that time had already reached $300 million.[68] The abolition of power-sharing agreements with Moscow, however, eliminated the legal basis for this claim.

The FMSF at Mayak was commissioned in December 2003, much later than originally anticipated. In part, the delay was caused by the reluctance of local construction firms to comply with the construction schedule because completing it would mean the loss of lucrative business.[69]

The interests of the oblasts adjacent to Chelyabinsk and Sverdlovsk Oblasts also may affect existing and future nuclear projects. The neighboring oblasts, especially the less economically stable ones, have already made attempts to bargain for compensation for radioactive contamination and risks they share with the oblasts that actually house nuclear facilities. For example, Kurgan Oblast demanded funds to clean up areas contaminated by the 1957 Mayak accident. Chelyabinsk Oblast, in turn, advanced claims for compensation for the deleterious environmental impact that is expected to result from chemical weapons elimination at the Shchuchye plant in Kurgan Oblast.

Factors That Threaten Nuclear Safety and Security in the Urals

Specific, current concerns about nuclear safety and security in Sverdlovsk and Chelyabinsk oblasts can be divided into several main categories.

---

[68] "Chelyabinskaya obl. obratilas k M. Kasyanovu s prosboy uchest interesy regiona," RIA RosBiznesKonsalting, July 2, 2001; Chelyabinsk oblast appeal to the President of the RF on the activities of PO Mayak and radioactive security in the Urals, April 27, 2000; "Russian Nuclear Security and the Clinton Administration's Fiscal Year 2000 Expanded Threat Reduction Initiative: A Summary of Congressional Action," RANSAC website, February 2000, <http://www.ransac.org>, pp. 17–18.

[69] As reported by Gen. Thomas Kuenning (ret.), Director of the Pentagon's Cooperative Threat Reduction Directorate, in his presentation at the CNS-sponsored workshop "Russian 'Nuclear Regionalism' and Challenges for U.S. Nonproliferation Assistance Programs," April 5, 2002, Washington, D.C.

Energy Problems

As mentioned above, energy shortages and power outages have rendered nuclear facilities in the Urals seriously vulnerable, from a safety and security perspective. Growing energy consumption by nearby industries will make the reliability of power supply to nuclear facilities, especially those housing operational nuclear reactors, even more critical.

Metal Theft

Due to the large number of metallurgical enterprises in the region and their profitability, it is easy to sell ferrous and non-ferrous scrap metal. Accordingly, metal theft has become very common. Nuclear enterprises, in particular, offer lucrative targets, despite being situated within closed cities.

There have been numerous reports of metal theft from nuclear facilities and closed cities, including by facility employees and guards.[70] Several incidents are worth noting:

- In July 1999 at VNIITF in Snezhinsk, workers stole several tons of highly radioactive aluminum plates from an "experimental area."[71]
- In October 1999, 10,000 meters of aluminum cable were cut from power lines near the closed city of Uralskiy (Sverdlovsk Oblast), leading to a power outage in the city and the nearby missile storage facility as well.[72]
- In August 2001, a track inspector prevented the theft of 200 meters of a railroad line leading to the Mayak industrial zone. The rails were already dismantled and ready for pick-up at the time of the discovery of the crime.[73]
- In March 2002, Mayak General Director Vitaliy Sadovnikov reported the theft of a sheet of steel weighing 150 metric tons, as well as continuing problems with thefts at Mayak during 2001-02.[74]

Despite the fact that metal and metal products are the number one item on the export list from both oblasts and bring in the most revenue to oblast budgets, the negative consequences of metal theft for the economy and security in the region

[70] Dmitry B. Zobkov, "Ukrali tonnu radioaktivnoy stali," *Kommersant*, December 18, 1999; "'Zvon' na svalke," *Ozerskiy vestnik*, No. 159, November 17, 1999, p. 2.
[71] Sergey Antonov, "Vernite nashu radiatsiyu," *Segodnya* online edition, July 3, 1999, <http://www.eastview.com>.
[72] "Vblizi poselka Uralskogo zloumyshlenniki srezali 10,000 metrov alyuminiyevogo provoda," Region-Inform, October 25, 1999; in Integrum Techno, <http://www.integrum.ru>.
[73] V. Riskin, "Ostanovili liternyy," *Chelyabinskiy rabochiy*, August 23, 2001.
[74] Report on the Mayak annual conference was compiled and distributed by Agentstvo Informatsionnogo Vzaimodeystviya (Agency for Information Cooperation, Ozersk), March 28, 2002.

undermine the benefits from metal exports. Some closed cities have prohibited scrap metal shops from their territory. However, local and regional attempts to curtail metal theft have had limited success to date. According to a petition to President Putin adopted by the Sverdlovsk legislature in May 2001, only nationwide efforts can stop illegal transactions involving scrap metal. The petition calls for a state monopoly in the acquisition and sale of scrap metal, strict enforcement of existing legislation, and the adoption of a federal law on this issue.[75]

General Crime Levels and Drug Abuse

At the October 2000 meeting of representatives from regional law enforcement agencies in Yekaterinburg, Federal Security Service (FSB) Director Nikolay Patrushev called the Urals federal district a "center of organized crime, drug abuse, and corruption."[76] Indeed, the crime level in the Urals is 20 percent higher than in Russia in general.[77] Although closed cities are still much better off than the rest of the region in terms of overall crime levels, the growth of the crime rate in these cities is also very high. In 2001, Ozersk registered a 3.4-fold increase in street crime, the bulk of which is thefts from cars and drug sales.[78]

Chelyabinsk Oblast borders Kazakhstan and serves as a route for drug smuggling from Central Asia to Russia and Europe. After the collapse of the Soviet Union, the border with Kazakhstan remains to a large degree "open" to drug traffickers. Chelyabinsk Oblast's efforts to close the border are not sufficient to solve the problem, even if a visa regime between the oblast and Kazakhstan is introduced, as mayors of some border cities have requested. Federal funding and tight law enforcement are needed to stop the drug trade and other illicit trafficking.[79] Reportedly, sales from illegal drugs in the Urals federal district total $1 billion.[80] Drugs easily find their way to Sverdlovsk Oblast via the Chelyabinsk and Kurgan Oblasts. Chelyabinsk and Sverdlovsk, along with Moscow, are the largest narcotics transit points in Russia.[81] Cheap, high-quality heroin is in

---

[75] "Putina poprosili zapretit sbor tsvetmeta," Region-Inform, May 24, 2001; in APN news, <http://www.apn.ru>.

[76] "Nikolay Patrushev: Uralskiy okrug—eto organizovannaya prestupnost, narkomaniya i korruptsiya," Strana.ru website , October 26, 2000, <http://www.strana.ru>.

[77] "Uroven prestupnosti na Urale prevyshayet srednerossiiskiye pokazateli," Izvestia.ru website, October 8, 2003, <http://main.izv.info>.

[78] "V tekushchem godu rezko—v 3.4 raza—vozrosla ulichnaya prestupnost," Ozersk.ru website, October 3, 2001, <http://www.ozersk.ru>.

[79] "Chelyabinskiye vlasti rasporyadilis 'zakryt granitsu dlya narkotikov'," Strana.ru website, June 27, 2001, <http://www.strana.ru>.

[80] Yekaterina Mineyeva, "Ural—odurmanennyy kray derzhavy," *Chelyabinskiy rabochiy*, June 21, 2001.

[81] Leonid Bukatin, "Seans igloukalyvaniya Rossii," *Novaya gazeta* online version, September 24, 2001, <http://www.NovayaGazeta.ru>.

abundant supply and contributes to high drug abuse rates in both oblasts. Most of the drugs originate in Afghanistan and Tajikistan.[82]

Drug abuse has spread to the closed nuclear cities as well. Ozersk is the leader in the growth of drug abuse among the closed nuclear cities in the Urals.[83] The local newspaper *Ozerskiy vestnik* reported that on several occasions even Mayak employees have been caught "under the influence."[84] The same newspaper reported that despite document checks at city entrance posts, many businessmen and mafia members from Yekaterinburg and Chelyabinsk regularly enter the city without difficulty.[85]

Existing contraband routes from Central Asia put the region at risk of becoming a venue for other illegal operations. The possibility of attempts by "rogue" states or terrorist groups to acquire nuclear material and technology or to commit a terrorist attack on one of the nuclear facilities cannot be ruled out. According to the head of the FSB in Chelyabinsk Oblast, an attempt to acquire the blueprints of one of the nuclear missile production facilities in the oblast was prevented in 1999. The arrested buyer was from "one of the Southeast Asian countries."[86] In 1998, the Chelyabinsk directorate of the FSB reported an attempted theft of nuclear material from one of the oblast's nuclear facilities.[87] In October 1999, the press service of the Russian MOD announced that Chechen leaders were preparing attacks on Russian nuclear installations, including nuclear power plants. It was reported that the Urals region, particularly Chelyabinsk Oblast, was a likely target.[88]

Inadequate Security

Despite the numerous programs and extensive international assistance aimed at addressing nuclear materials physical protection and accountancy at Minatom facilities, the upgrades provided were not comprehensive. Moreover, some facilities did not receive any foreign assistance because access to them has been denied. The U.S. Department of Energy MPC&A Program has been waiting for years to receive permission to visit Minatom serial production plants in order to

---

[82] Vyacheslav Yakovlev, "Cherez nashy oblast narkobarony napravlyayut karavany otravy iz Azii v Yevropu," *Chelyabinskaya oblast* on-line version, No. 5, August 30, 2000, <http://www.chelinform.ru>.

[83] Interviews with residents of Ozersk, Snezhinsk, and Novouralsk, November 2000.

[84] Yelena Vyatkina, "Opasnyye lyudi na opasnom proizvodstve," *Ozerskiy Vestnik*, No. 167, December 1, 1999, pp. 6–7.

[85] "Voruyut bukvalno vsye," *Ozerskiy Vestnik*, No. 145-146, October 23, 1999, p. 22.

[86] Aleksandr Bragin, "Ya—za diktaturu zakona," *Yuzhnouralskaya panorama*, May 25, 2000, <http://www.chelpress.ru>; Aleksandr Bragin, "Spetssluzhby—vne politiki," *Chelyabinskiy rabochiy*, July 28, 2000, <http://www.chelpress.ru>.

[87] Yevgeniy Tkachenko, "FSB Agents Prevent Theft of Nuclear Materials," ITAR-TASS, December 18, 1998.

[88] Mikhail Tolpegin, "Chechnya Prepares Nuclear Strike," *Segodnya*, October 13, 1999, p. 3.

assess security levels and propose upgrades. Two of these facilities are in Lesnoy and Trekhgornyy in the Urals. The arrest of a Chechen arms dealer in Yekaterinburg with a valid pass to the closed city Lesnoy, home of a nuclear warhead production facility, points to the shortcomings of the existing security system.[89] Regional authorities continue to voice their concerns over the level of physical protection of nuclear facilities in the Urals and criticize Minatom and Moscow for slow progress and reductions in the number of nuclear facility guards.[90]

The Role of the Presidential Envoy

After his appointment to the Urals federal district in May 2000, Presidential Envoy Latyshev concentrated on harmonizing regional laws with federal norms, replacing key federal employees in the district who had been hand-fed by the governors, and wrestling for the support of the Urals business elite. With only administrative rather than financial levers at his disposal, it was not an easy task, especially in Sverdlovsk Oblast. Latyshev started by establishing control over the regional police, security services, and other law enforcement agencies. He created a district security council that includes the heads of these agencies in the region. Yet his involvement in nuclear issues at the beginning of his appointment was minimal. Gradually, Latyshev was able to gain political clout in the district. The events of September 11th have given an additional thrust for Latyshev to be more actively involved in nuclear issues in the Urals. He has held regular meetings on nuclear industry development in the Urals, nuclear safety and security, antiterrorist measures, and the coordination of various federal and regional law enforcement and emergency response agencies in the district.[91] Along with the regional governors, Latyshev met with senior government and Minatom officials during their visits to the region's nuclear facilities.

Conclusion

Traditionally, regional nuclear facilities have strongly associated themselves with Minatom, Moscow, and to a much lesser degree with attendant regional

---

[89] Sergey Avdeyev, "Chechentsy poluchili dostup k yadernym boegolovkam," *Izvestiya* online edition, March 21, 2002, <http://www.izvestia.ru>.

[90] "Sverdlovskiy gubernator ozabochen sostoyaniyem okhrany yadernykh obyektov," RIA Novosti–Ural, October 29, 2002, <http://ural.rian.ru>.

[91] See for example, Eduard Puzyrev, "Putin Envoy Calls For Tighter Security At Russian Nuclear Power Stations," RIA-Novosti, February 12, 2002; in RANSAC Nuclear News, February 15, 2002. Also see, "Putin Envoy in Urals Calls for Better Antiterrorist Protection of Nuclear Plants," Nuclear.ru website, May 17, 2002, <http://www.nuclear.ru>; in FBIS Document CEP20020522000088.

administrations. However, since the Soviet Union collapse, they have increasingly found themselves dependent on the region, whether for a permit to build a new facility or start a new operation, a steady supply of power and water, or fulfillment of contracts by oblast industries involved in domestic and international projects. The regions have been and remain a source of supplemental emergency funding; regional leaders can bolster public support for, or silence opposition to, nuclear facility plans; and regional authorities can lobby on the facilities' behalf at the federal level. Regional governments continue to pursue bilateral agreements with Minatom, which is more effective in solving nuclear facilities and regional concerns or economic interests than dealing with the federal government as a whole or the parliament. This trend has persisted, despite the stabilization of federal funding for Russian nuclear weapons production facilities and Putin's recentralization campaign. Nuclear facilities and closed cities in the Urals still find it necessary to be represented at the oblast level and get involved in the political and economic processes in their respective oblasts. In Chelyabinsk Oblast, all three Minatom closed cities have deputies in the oblast Duma.

Governors also consider closed cities an important part of the regional electorate, and the most disciplined and active, for that matter. The frequency of the governors' visits to closed cities and attention to their needs usually increases during election campaigns. In November 2000, on the eve of his re-election, Governor Sumin sent a letter to President Putin calling for the renewal of the construction of the South Urals NPP. Mayak Director Sadovnikov, in turn, served as an official representative of Governor Sumin in the fall 2000 elections. According to some reports, "the governor's new crusade for the South Urals NPP is the price for a pre-election deal with 'rich Mayak'."[92] Moreover, the directors of income-generating nuclear facilities themselves often are seen as a part of the regional business and political elite.

The redistribution of tax revenues in favor of the federal government in 2001-2002 have made economic considerations even more important for the regions, as they lost other bargaining chips vis-à-vis the federal center. On the other hand, if the regions do not receive the promised benefits from hosting nuclear facilities or servicing their needs, their support for Minatom activities might decline and endanger the existing balance of relations in that area. The shifting attitude towards the import of spent fuel and its reprocessing has already demonstrated the economic underpinning of regional nuclear politics.

At the same time, political support for Minatom will hardly disappear completely because the two oblasts are likely to remain committed to the expansion of nuclear power generation and construction of new civilian reactors, as they expect to benefit from more reliable, cheaper energy and from contracts for their construction, including for the oblast machine-building industry. Sverdlovsk and Chelyabinsk Oblasts also most likely will remain vocal proponents of contracts for NPP construction abroad, since many regional enterprises (including private

---

92 "Soglasen li rossiyskiy nalogoplatelshchik vylozhit iz karmana 1,5 milliarda dollarov za YuUAES?" *Delovoy Ural* online edition, November 23, 2000, <http://www.infoural.ru>.

machine-building companies such as UralMash) participate in them. According to Deputy Minister of Atomic Energy Bulat Nigmatulin, the contracts for Russian machine-building enterprises involved in the production of equipment for foreign power plants could reach $1.4 billion by 2007. This includes six reactors for China, Iran, and India.[93] Fears of U.S. concerns over nuclear cooperation with these countries are outweighed by the expected income to the region. In addition, anti-American sentiments in these two oblasts are further reinforced by the losses incurred by steel producers in the Urals from restrictions on imports established by the U.S. government under the Bush administration. Regional political efforts to influence national policy on this issue will be matched by the efforts of financial-industrial groups that operate in these industries.

The new trend toward recentralizing the flow of finances from profitable nuclear-sector activities, including those under international programs, in favor of Moscow and Minatom could have a serious negative impact on those nuclear facilities and closed cities that previously generated considerable revenues from these programs. The loss of revenues will affect social conditions, payment of wages, and economic diversification. At the same time, the regions hardly will be able to continue to act as "emergency backup" for closed cities that depend wholly or partially on federal funding. Any shortfall in federal funding will put the security and stability of the nuclear sector at risk, similar to what took place in the early 1990s. These concerns are particularly acute in light of the on-going discussions of Minatom restructuring that could lead to the split of its power production, nuclear fuel, and defense complexes and incorporation of them into three or more government agencies. In addition, there are plans for the elimination of the special financing mechanism for ZATOs. In this case, the closed nuclear cities will have to operate in a much more austere financial environment, and thus will be more dependent on oblast budgets.[94]

In sum, the Chelyabinsk and Sverdlovsk Oblasts will continue to play a central role in the Russian nuclear complex, including in the defense sector. However, the long-term orientation of the Urals nuclear facilities toward large defense orders and federal funding, slow progress in the diversification of economic activities at facilities and in closed cities, and their weak incorporation into regional economies will continue to pose challenges to managing these facilities, personnel, and assets in the future at all levels of government. The role of local and regional factors in their current and future operation will most likely continue to grow, particularly if special federal funding to closed cities is cancelled and the nuclear defense complex embarks on downsizing and conversion in earnest.

---

[93] Ibid.

[94] Some reports suggest that ZATOs will lose their privileged financial status and direct federal funding in 2005. See the discussion of this issue in Elena Mazanova, "Pogasit li Ilkayev inflyatsiyu?" *Gorodskoy Kurer* on-line edition, January 1, 2004, <http://courier.sarov.ru>.

Chapter 9

Nuclear Issues in the Siberian Federal Okrug

Dmitry Kovchegin

The Siberian Federal Okrug (district) represents a unique case of nuclear decentralization and regionalist tendencies in Russia.[1] It has been home to some of the most flagrant episodes of nuclear assertiveness. Ambitious regional leaders have also played the nuclear card in their political gambits at the federal level. The former governor of Krasnoyarsk Kray, for example, went so far as to threaten, though mostly for political bravado, to commandeer the Strategic Rocket Forces (SRF) deployed on his territory if the federal authorities failed to solve the financial problems of his SRF unit. At the same time, with its high concentration of nuclear facilities, the okrug has seen the considerable influence that the nuclear lobby can wield over regional political and economic affairs. For example, one of the world's largest nuclear complexes, the Siberian Chemical Combine (SKhK) located in Tomsk Oblast, has turned its attention inward to dominate the local political scene.

But the Siberian district traditionally has assumed an inconspicuous posture in the national debate over federalism. Notwithstanding its significant federal footprint, covering 30 percent of Russia and accounting for 14.3 percent of the national population, the region has not been prone to embracing strong nationalist or separatist sentiments.[2] In general, the constituent administrations of the okrug have been noteworthy for their general accommodation of federal interests. For

---

[1] The Siberian Federal Okrug comprises the Altay, Buryatiya, Tuva, and Khakassiya republics; Altay and Krasnoyarsk Krays; Irkustk, Kemerovo, Novosibirsk, Omsk, Tomsk and Chita Oblasts; and Agin- Buryat, Taymyr (Dolgano-Nenetskiy), Ust-Orda Buryat, and Evenki Autonomous Okrugs. The center of the federal okrug is Novosibirsk.

[2] *Regiony Rossiy, 1999* (Moscow: Russian State Committee for Statistics, 1999). The autonomous regions are part of the larger administrative-territorial formations. This is quite a confusing picture. This situation led to the conclusion of trilateral agreements on delimiting authority between federation constituencies and the center. However, the situation has recently begun to change. A trend toward simpler relations between the center and the regions is reflected by the efforts Ust-Orda Buryat Autonomous Okrug to closer relations with Irkutsk Oblast, of which it is technically a part. The final goal of the constituencies is administrative unification.

instance, local administrations reacted calmly to President Vladimir Putin's 2000 reform that created seven federal districts headed by newly appointed presidential plenipotentiaries. According to new Siberian Governor Leonid Drachevskiy, the okrug's warm reception enabled him to find his niche fairly quickly so that he did not interfere with the functions of the territorial leaders and did not disrupt—but, in fact, strengthened—the vertical chain of command. As a result, smooth and constructive working relations have been established with the heads of other federation constituencies. These relations have improved the effectiveness of the local authorities.[3]

How do we reconcile the seemingly conflicting postures of nuclear activism with the okrug's generally passive stance on federalism? Moreover, why has nuclear decentralization assumed different faces among the constituent parts of the okrug? Why, for example, has Krasnoyarsk Kray been home to provocative initiatives directed against the federal center, while in Tomsk, the significant influence of the nuclear lobby has taken on a subtle but nonetheless formidable regional focus?

This chapter addresses these issues by analyzing the political economy of the two dominant nuclear territories in the Siberian Federal Okrug: Krasnoyarsk Kray and Tomsk Oblast. It begins with an overview of the defining political and economic characteristics of the okrug, including a description of the nuclear assets across the territory. The second section reviews the nuclear landscape in Krasnoyarsk Kray, and analyzes the specific political conditions that motivated former Governor Aleksandr Lebed to brandish the nuclear card in relations with the federal government. Section three reviews the nuclear infrastructure in Tomsk Oblast, and traces the apparent *quid pro quo* between the local administration and nuclear weapons complex that forms the basis of a distinctively cooperative face of nuclear activism. The final section develops some of the implications of these two forms of local activity for nuclear regionalism in Russia.

General Characteristics of the Siberian Federal Okrug

Political Economy of the Okrug

The economic fault lines of the Siberian Federal Okrug define its political character. The okrug is noteworthy for its large natural resource endowment, particularly in oil, gas, and timber. The local industry generally complements this profile, dominated by automotive manufacturing, metallurgy, fuel, chemical, timber, and power generation. This combination has given rise to a conspicuous export profile, with the region maintaining an import/export ratio of 30:70 percent,

---

[3] According to Dracheviskiy, "We don't touch issues that fall under the jurisdiction of the governors and don't step on their territories." See "Budet li mogushchestvo Sibiri polpredom prerastat?" *Rossiyskaya gazeta*, September 30, 2000.

as compared to the all-Russian ratio of 60:40 percent.[4] This profile also lured large financial-industrial groups (FIGs) to the okrug, such as Interros (Norilsk Nikel), RAO EES, Sibirskiy Alyumniy, and Yukos Oil. While the focus of business activity in the mid-1990s generally served narrow commercial interests (i.e., the acquisition of profitable factories, the organization of regional financial outlets, cooperation with representatives of the local elite, a search for lobbyists among local high-level bureaucrats to secure local market niches, and the liquidation of rival regional companies), the FIGs have gradually acquired an appreciable stake in regional politics.[5] For example, then-Norilsk Nikel General Director Aleksandr Khloponin mounted a successful 2001 run for governor in Taymyr Autonomous Okrug[6] and eventually became governor in Krasnoyarsk Kray as a result of the special election in the fall of 2002. Another example of the political penetration of big business took place in Krasnoyarsk Kray, where local oligarch Anatoly Bykov has acquired the status of political boss over the past few years.

One of the most important entities in the region is the Siberian Agreement (Mezhregionalnaya Assotsiatsiya Sibirskoe Soglashenie, or MASS), an inter-regional association for economic cooperation that was formed in 1990. It is comprised of all of the federal constituencies within the Siberian Federal Okrug. MASS's main tasks are to: facilitate intra-regional material, financial, and intellectual cooperation in order to further socio-economic development and create a higher standard of living for the population; to establish a single economic, legal, industrial, and informational space among its membership; and to draft proposals concerning international and foreign trade ties and the implementation of regional investment programs and projects. Since its inception, MASS has influenced the federal center on questions regarding horizontal ties among the regions. A clear example of this occurred at a 1999 okrug budget council review meeting attended by President Putin. The council capitalized on this opportunity to coordinate the redistribution of tax revenues within the budget before it was submitted to the State Duma.[7] That the Siberian Accord is represented by 68 deputies in the State Duma speaks to its national prominence.[8]

President Putin's creation of federal okrugs was not initially welcomed by MASS. The MASS leadership considered possible dissolution of the association, as responsibilities held by the Siberian Federal Okrug and MASS are generally redundant. Yet, Presidential Representative Drachevskiy claimed that "the presidential representative for the Siberian Federal Okrug and the Siberian Agreement interregional association are complementary structures and are not mutually exclusive." In his opinion, MASS played a critical supporting role for

---

[4] Calculations based on data from the Russian State Committee for Statistics.
[5] Andrey Serenko, "Novaya regionalnaya politika," *NG-regiony*, No. 6, April 11, 2000.
[6] Russian Federation Central Election Commission website, <http://www.fci.ru>.
[7] Yuriy Trigubovich, "Nuzhna li Rossii Sibir?" *NG-Regiony*, No. 3, February 8, 2000.
[8] Interview with State Duma Deputy Aleksandr Fomin, January 11, 2001, <http://www.sfo.nsk.su/fomin.htm>.

both the legislative and executive branches within the okrug.[9] Federal authorities responded by treading lightly so as not to infringe on MASS. Consequently, both levels of the administration have worked closely ever since.

The Nuclear Structure of the Siberian Federal Okrug

The Siberian Federal Okrug is marked by a high concentration of nuclear assets (see Tables 9.1-9.3). There are approximately 20 nuclear facilities located in the okrug that are under the administrative control of either the Ministry of Defense (MOD) or the Ministry of Atomic Energy (Minatom). The characteristic feature of the majority of Minatom enterprises within the territory is that they are involved in nuclear fuel cycle production activity and now, to a large extent, are servicing the civilian sector (although earlier they were involved in both defense and civilian production). Because the standard of living of the employees at these nuclear fuel cycle factories is typically higher than in the region as a whole, the nuclear presence has elevated socio-economic conditions in the immediately surrounding areas of the okrug. This factor, in turn, has generally enhanced the ability of the nuclear industrial enterprises to influence local politics.

Table 9.1 Nuclear Facilities in the Siberian Federal District: Ministry of Defense Facilities

Location	Name	Comments
Aleysk (Altay Kray)	Former ICBM Base	Formerly 30 RS-20 (SS-18)[10]; officially deactivated in May 2001
Barnaul (Altay Kray)	ICBM Base	36 RS-12M (SS-25)
Drovyanaya (Chita Oblast)	ICBM Base	18 RS-12M (SS-25)
Yasnaya (Chita Oblast)	Former ICBM Base	All missiles here were destroyed by 1994.
Irkutsk	ICBM Base	36 RS-12M (SS-25)
Kansk (Krasnoyarsk Kray)	ICBM Base	45 RS-12M (SS-25)
Krasnoyarsk	ICBM Base	12 SS-24 (rail-mobile launchers)
Yeniseysk (Krasnoyarsk Kray)	Former ballistic missile early warning system radar station	Declared to be in violation of the ABM Treaty and dismantled in 1989
Uzhur (Krasnoyarsk Kray)	ICBM Base	52 SS-18
Novosibirsk	ICBM Base	45 SS-25

[9] "Obyedinyay i vlastvuy," *Vek*, November 24, 2000.

[10] Nikolay Petrov, "Voyna s neveroyatnym protivnikom," *Kommersant-Vlast*, No. 29, July 25, 2000, pp. 16–20.

Table 9.2 Nuclear Facilities in the Siberian Federal District: Ministry of Atomic Energy Facilities

Location	Name	Comments
Krasnokamensk (Chita Oblast)	Priargunskiy Mining and Chemical Production Association	Mines, enriches, and processes uranium ore. The final product is natural-enrichment uranium oxide.
Angarsk (Irkutsk Oblast)	GUP Angarsk Electrolytic Chemical Combine	Produces uranium hexafluoride and enriches uranium (up to 5 percent U-235). More than 50 percent of the production of the factory is for export.
Irkutsk	Radon Special Enterprise	Stores radioactive waste
Zelenogorsk (Krasnoyarsk Kray)	Electrochemical Plant (EKhZ)	Uranium enrichment facility. Produces low-enriched uranium and participates in the HEU-LEU deal
	Mining and Chemical Combine (GKhK)	Produces plutonium; reprocesses fuel from production reactors; stores spent nuclear fuel. A spent nuclear fuel reprocessing plant is planned.
Novosibirsk	AO Novosibirsk Chemical Concentrate Plant (NZKhK)	Produces fuel for Russian and foreign nuclear power plants reactors (VVER-1000 and VVER-400 reactors); HEU fuel for research reactors and fuel for plutonium production reactors
	Radon Special Enterprise	Stores radioactive waste
	Sever Production Association	Development and production of equipment for nuclear weapons.
Seversk (Tomsk Oblast)	Siberian Chemical Combine (SKhK)	Production in all areas of the nuclear fuel cycle
Krasnoyarsk	Krasnoyarsk Chemical and Metallurgical Plant	Produces ceramic UO_2 and lithium hydroxide

Table 9.3 Nuclear Facilities in the Siberian Federal District: Other Nuclear Facilties

Location	Name	Comments
Tomsk	Scientific Research Institute of Nuclear Physics of Tomsk Polytechnic University (under the Ministry of Education)	Conducts research and training in nuclear physics, accelerator technologies, and electronics; houses IRT-T training and nuclear research reactor
Krasnoyarsk	Krasnoyarsk Chemical and Metallurgical Plant	Produces ceramic UO_2 and lithium hydroxide

After an initial decline, production and socio-economic conditions began to improve within Minatom's closed cities, or ZATOs, when the directors of the enterprises and political leadership of the corresponding regions realized that their future rested not with large-scale conversion, but with a slight reorientation of their nuclear activities. According to State Duma Deputy Valeriy Zubov, who represents the electoral district surrounding the closed city of Zheleznogorsk, which is dominated by the Mining and Chemical Combine (GKhK), "The best direction for the GKhK is the processing of spent nuclear fuel."[11]

Thus far, most of the region's nuclear enterprises have not been subject to drastic personnel cuts. However, this situation could change in the near future after the planned shutdown of the plutonium production reactors in Seversk and Zheleznogorsk. It is expected that as a result of this process several thousand nuclear employees will have to look for other jobs. This is a substantial share of the facilities' personnel. Keeping the social situation stable will present a significant challenge for Minatom and ZATO (closed city) authorities.

The rise of worldwide and domestic terrorist activity poses another major threat for the region's nuclear facilities, their surrounding communities, and global society as a whole. Two major scenarios should be considered: 1) sabotage against a nuclear facility with the goal of releasing hazardous substances to contaminate both surrounding and (in the worst case) distant territories; and 2) the theft of weapons-usable material for construction of a dirty bomb. These extreme cases of terrorism cannot be wholly protected against given the nuclear facilities' current capabilities, which rely only on certain technical means of physical protection plus a limited guard force. That is why region- or even federation-wide cooperative efforts by law enforcement agencies, regional and local authorities, and nuclear officials are needed to cope with this threat.[12] Law enforcement authorities have

[11] Interview with Valeriy Zubov, January 5, 2001.

[12] Dmitry Kovchegin, "Approaches to Design Basis Threat in Russia in the Context of Significant Increase of Terrorist Activity," in *Proceedings of the 44th Annual Meeting of the Institute for Nuclear Materials Management, Phoenix, Arizona, July 13-17, 2003* (Northbrook, Illinois: INMM, 2003).

reacted to recent terrorist alerts by increasing the concentration of troops around nuclear facilities.

There are also a number of cooperative U.S.-Russian nonproliferation programs being implemented at the region's nuclear facilities. The most important of them include:

- *Material protection, control, and accounting* (MPC&A). Through this program, the U.S. Department of Energy provides funding for MPC&A equipment and training at facilities containing weapon-usable nuclear materials.
- *Highly enriched uranium (HEU) to low-enriched uranium (LEU) deal*. In the framework of this program, Russian blends down 500 tons of HEU declared excess to military purposes to sell it for use as fuel in U.S. nuclear power plants. This deal serves as a significant source of income for the Russian enterprises involved.
- *Plutonium production reactor shutdown*. This program is aimed at stopping additional production of weapons-grade plutonium in the reactors at Seversk and Zheleznogorsk. Since these reactors also serve as energy suppliers for local communities, alternative energy sources (fossil fuel power plants) will be provided to compensate losses of nuclear energy production capacities.

Krasnoyarsk Kray: A Microcosm of Russia

Krasnoyarsk Kray constitutes 14 percent of Russia's territory, but only 2 percent of its population (with approximately 3,075,600 people), mostly living in cities. The kray also subsumes the Taymyr and Evenki Autonomous Okrugs, which pose center-regional issues within this federation's constituency and create uneven representation among the constituent parts in the federal legislature.[13] Krasnoyarsk Kray inherited a diversified economy dominated by raw material exports and the remnants of the decaying military-industrial complex, which is also true for Russia in general. A significant role in the kray is played by financial-industrial groups that operate at both the regional and national levels. With respect to a variety of socio-economic indicators—such as changes in industrial productivity, unemployment, income, and cost of living—Krasnoyarsk Kray mirrors national averages, albeit on a smaller scale. In addition to the GKhK in Zheleznogorsk (mentioned above), the kray's nuclear infrastructure includes the Electrochemical

---

[13] The autonomous districts that are incorporated in Krasnoyarsk Kray are also considered federation constituencies, according to the constitution. This allows them to have two representatives in the Federation Council of the federation. Two more directly represent Krasnoyarsk Kray. As for deputies of the State Duma, Krasnoyarsk Kray is represented by four deputies, and the autonomous regions by one apiece. The number of voters in the Taymyr Autonomous Okrug is 26,000, the Evenki Autonomous Okrug has 12,000, and Krasnoyarsk Kray has more than 2 million.

Plant (EKhZ) in Zelenogorsk and four MOD facilities—the Kansk, Krasnoyarsk, and Uzhur ICBM bases and the early warning radar station in Yeniseysk.

The Political Scene in Krasnoyarsk Kray[14]

Until his abrupt death in the spring of 2002, the key political figure in Krasnoyarsk Kray with respect to nuclear regionalism was Governor Lebed.[15] Lebed gained national prominence by defending the Russian minority in an armed conflict in the Trans-Dniester Republic. In January 1996, he launched a campaign for the presidency as an independent candidate. He polled third in the first round, with 14.7 percent of the national vote, forcing a run-off that ultimately secured the reelection of the incumbent, Boris Yeltsin. From June to October 1996, Lebed held the post of secretary of the Russian Security Council, during which time he signed the Khasavyurt peace agreement with Chechen separatists. Lebed also gained notoriety for issuing unsubstantiated claims concerning the possible loss of "suitcase" nuclear weapons by Russia's Special Forces.

On May 17, 1998, Lebed won the second round of elections for governor of Krasnoyarsk Kray, with 57 percent of the vote, while his opponent, incumbent Governor Valeriy Zubov, garnered only 38 percent. Becoming governor, however, was not Lebed's primary political objective at the time. Rather, he viewed the kray election as a springboard for a second assault on the Russian presidency that, he presumed, would be held shortly thereafter due to President Yeltsin's early retirement for nagging health reasons. Based on the assumption that he would occupy the post only briefly, Lebed focused attention immediately on preparing to run in the upcoming presidential election.[16] This led him to challenge federal officials at all levels, trumpeting the cause of regional autonomy. Accordingly, he leveled successive criticisms of the center's policies toward the regions and demanded more political autonomy for the federation's constituent parts. Lebed flatly dismissed accusations that he was fomenting separatism, which at that time could have hurt his presidential candidacy. He stated that "Separatism in the middle of Russia is categorically counterproductive. I am interested in a single, unified Russia with Krasnoyarsk Kray as an integral part."[17]

Yet, with Vladimir Putin's ascendance on the national political scene from summer 1999 to winter 2000, first as prime minister and then as the frontrunner to succeed President Yeltsin, Governor Lebed clearly understood that he no longer had a realistic chance to become president. He re-directed his efforts to working

---

[14] Since Aleksandr Lebed is a key figure in the nuclear regionalism question and an analysis of his policies provides insight into Russian nuclear issues more generally, this section is mainly devoted to General Lebed.

[15] For more information, see Nikolay Petrov, *Aleksandr Lebed v Krasnoyarskom kraye* (Moscow: Carnegie Center, 1999).

[16] Igor Oleynik, "Proriv v gubernatory Krasnoyarskogo kraya," in *Aleksandr Lebed i vlast* (Krasnoyarsk, 1998).

[17] Announcement of Aleksandr Lebed in New York, July 4, 1998.

with the Putin administration on key national issues. During this period, Lebed did not exploit ambiguity in the constitution that would have allowed him to exercise autonomy in the foreign policy sphere. In particular, he did not attempt to work at cross-purposes in areas of joint jurisdiction, even on those foreign trade and interregional issues that were legally under the exclusive competence of the kray administration. According to foreign ministry officials, there were no discrepancies between the federal center and the kray administration on issues of international political and economic affairs. To the extent that there were any problems, they were related to administrative delays in submitting kray international agreements for federal approval.[18]

Rebuffed at the federal level, Lebed shifted attention to redressing the socioeconomic problems confronting Krasnoyarsk Kray and to sustaining his regional power base until a future opportunity opened to regain his national stature. As Lebed turned inward to tackle regional affairs, he confronted political challenges on two fronts. Within the region, the two most powerful lobbies were the aluminum group, headed by Bykov, a regional oligarch and shrewd political operator[19]; and the Norilsk group, based on RAO Norilsk Nickel controlled by the ONEKSIM-Interros financial group. Both groups derived respective influence from their concentration of significant economic capital and their weighty role in forming the regional budget. Complicating this situation was the fact that federal-level political parties were largely inactive in the kray between national elections, which, in turn, allowed regional organizations to compete successfully against them in local elections.[20] Yet these industrial atavisms continued to dominate the political scene within the closed cities. For example, GKhK representatives won five of the 24 seats in the Zheleznogorsk city council during the January 2001

---

[18] Interview with an official at the Department of Relations Between Federation Constituencies and RF Parliamentary Organizations of the Russian Ministry of Foreign Affairs, December 27, 2000. In accordance with the Constitution of the Russian Federation (articles 71-72) the Russian Federation has jurisdiction over foreign policy and international relations, international agreements, and also foreign trade relations. The Russian Federation and federation constituencies share authority in coordinating the constituencies' international political and economic ties, and also the fulfillment of international agreements Russia has entered. At the same time, according to Article 14 of the charter of Krasnoyarsk Kray, international economic relations and international and interregional relations are under the jurisdiction of the kray. The legislative basis of international economic relations of the federation constituencies is the law "On coordinating international and foreign trade relations of federation constituencies," which entered into force on January 16, 1999.

[19] He was eventually arrested on accusations ranging from illegal actions of a financial-economic nature to assassination. In June 2002, he was released on probation.

[20] This is due largely to the fact that during the 1990s, the main role in the kray was played by a group of industrial "generals," who have now lost their influence due to reforms. For example, in the elections for the city council of Krasnoyarsk the first two positions were won by the "Bloc of Anatoly Bykov" and the bloc "For Krasnoyarsk," headed by the reelected mayor of Krasnoyarsk, Pyotr Pimashkov. Nikolay Petrov, *Aleksandr Lebed v Krasnoyarskom kraye* (Moscow: Carnegie Center, 1999), p. 72.

elections.[21] Ultimately, however, Lebed was not successful in his struggle with the influential groups in the kray and, in the end, had to turn to federal authorities for help, reversing his rhetoric from opposition to the federal center to become its main supporter within the region.

Current Krasnoyarsk Kray Governor Aleksandr Khloponin was elected in the fall of 2002 in the special election after General Lebed's death in a helicopter crash. From the point of view of center-region relations, the situation had significantly changed by this point. Vladimir Putin's reforms of center-region relations were nearly finished, and Krasnoyarsk Kray had lost its donor status. This led to a situation in which every candidate, in order to have good chances for election and further smooth governance, had to be loyal to federal authorities. Since the elections took place, no significant controversy has emerged between Krasnoyarsk Kray and federal authorities. In addition to serving as regional governor, Khloponin was selected to be a member of the Supreme Council of *Edinstvo (Unity)*—a major Russian party known for its implicit support of President Putin.

Governor Khloponin is a strong supporter of economic decentralization. However, his position does call for any kind of separatism. Indeed, his stance is supported by President Putin. Recently, Khloponin has asserted the need for regional enlargement on the basis of economic logic, while at the same time speaking in support of greater economic freedom for the regions since "it's impossible to develop the economy from Moscow."[22]

The Nuclear Infrastructure of Krasnoyarsk Kray

As noted above, the nuclear establishment in Krasnoyarsk Kray is dominated by two massive complexes: the EKhZ in Zelenogorsk (formerly Krasnoyarsk-45) and GKhK in Zheleznogorsk (formerly Krasnoyarsk-26). The present and future of these enterprises is determined to a large extent by their assignment and respective technologies used.

The main activity of EKhZ since the mid-1950s had been the production of HEU for weapons. In 1992, the plant shifted its focus to the production of LEU for nuclear power reactors. At present, it is also a participant in the U.S.-Russian HEU-LEU agreement, charged with blending down uranium extracted from nuclear weapons. In the 1990s, several successful non-military production lines were developed at the facility, and over 10,000 people have remained employed at the plant since the Soviet collapse.

During the Soviet period, the main purpose of GKhK was production of weapons-grade plutonium. Two out of its three plutonium-production reactors, however, were shut down in 1992. The third reactor remains open and produces

[21] *Gorodskoy Novosti* (Zheleznogorsk), January 18, 2001.

[22] Svetlana Babaeva, "Aleksandr Khloponin: 'Nel'zya podnimat' ekonomiku iz Moskvy'," *Izvestiya*, December 25, 2003.

heat and electricity for the city of Zheleznogorsk, although it is scheduled for shutdown in a few years, when alternative energy production capacity becomes available. Currently, the only stable source of income for GKhK comes from fees for the storage of spent VVER-1000 fuel from reactors in Russia and Russian-built foreign reactors. A wet storage facility with a capacity of up to 6,000 tons of spent fuel is used for this activity. There are plans to increase significantly the storage capacities at GKhK. The wet storage pool will be modified to accommodate up to 9,000 tons of spent fuel and a dry storage facility for 33,000 tons of spent fuel will be built.[23] No reprocessing is possible in the foreseeable future.

Both the GKhK and EKhZ complexes participate in joint U.S.-Russian programs to decrease the threat of nuclear weapons proliferation. For instance, the U.S. DOE program aimed at improving the nuclear MPC&A system began at both facilities in 1996 in accordance with a decision of the Gore-Chernomyrdin Commission. Implementation of the program continues to this day. Zheleznogorsk was also one of the three cities selected for implementation of the DOE-sponsored Nuclear Cities Initiative (NCI). This program began in September 1998 with the primary objective of creating new jobs in the civilian sphere for Russian nuclear specialists leaving the weapons program. In November 1999, the NCI-funded International Center for Development opened in Zheleznogorsk. The center's programmatic goals include: 1) providing services to expedite the conversion of production facilities, stimulating new entrepreneurs, and developing municipal projects; 2) drafting a strategic development plan for the city and business plans for converted enterprises; and 3) providing information and technical support for a variety of projects and programs implemented via U.S. assistance programs. Unfortunately, as of early 2004, the future of this program is unclear. It was initially signed for a five-year term, but was not renewed because of U.S.-Russian controversy over the liability issue. However, a number of projects initiated under NCI are still underway.

In addition, Zheleznogorsk has been a direct beneficiary of Russia's fiscal reforms. In response to tax breaks for businesses registered in the closed cities introduced by new Russian legislation, the Zheleznogorsk city council in March 1998 formed an investment zone in the closed city. The statute stipulated that ventures registered within the zone were entitled to significant tax benefits. In contrast with some other closed cities, Zheleznogorsk did not abuse its tax privileges and later was rewarded for its "good behavior." In accordance with federal budget laws implemented in 2000, the nuclear closed cities, excluding Sarov and Snezhinsk, no longer were permitted to extend federal tax benefits to commercial enterprises operating on their territory. This amendment was passed in response to the rampant abuse of earlier offshore tax privileges. In exchange, closed cities that demonstrate diligence in collecting and remitting federal taxes are

[23] Vladimir Korotkevich and Evgeny Kudryavtsev, *Tehnologiya I bezopasnost' obrasheniya s obluchennym yadernem toplivom v Rossiiskoi Federatsii* (*Technology and Safety of Spent Fuel Management in Russian Federation*). Byulleten' po atomnoi energii (Atomic Energy Bulletin), December 2002.

eligible to receive additional central financing. Zheleznogorsk has been designated as one of these cities, and has been allowed to receive significant financing from the federal budget.[24] Consequently, Zheleznogorsk has become heavily dependent on Moscow and has not looked to the Krasnoyarsk Kray administration for fiscal relief.[25]

At the same time, however, the relative amount of taxes received from enterprises and activities, unrelated to the nuclear complex in Zheleznogorsk, has risen substantially in the last few years.[26] This is mainly the result of the growing number of new enterprises that have been hived-off of existing facilities and financed by Minatom. This, in turn, has extended Minatom's influence in the city beyond the GKhK complex.

A major conversion project being realized in Zheleznogorsk is the creation of a plant to produce silicon for semi-conductors. This project was approved by governmental decree in June 1996 and is being implemented jointly by several ministries, regional authorities as well as other enterprises, and scientific establishments in the Siberian Federal Okrug.[27]

Krasnoyarsk Kray and the Effects of Nuclear Decentralization

Lebed's election as governor of Krasnoyarsk Kray coincided with the country's deepening economic crisis in 1998. Upon assuming leadership of the kray, Lebed was determined to use the worsening national situation to his advantage and to protect against a regional backlash. Accordingly, prior to the federal government's decision to devalue the currency in August, Governor Lebed sent an open letter to then-Russian Prime Minister Sergey Kiriyenko demanding that the federal government meet its fiscal obligations to the kray. In it, Lebed stated:

> Officers of the Uzhur Missile Division haven't received monetary compensation for five months, and their wives are storming the headquarters of the association. Sergey Vladilenovich, this is a serious condition, and the officers are in serious need. And I'm seriously thinking about putting the division under the jurisdiction of the kray. We, the people of Krasnoyarsk, are not yet rich, but in exchange for the status of nuclear kray, we will feed the division, though by doing this, we will become, like India and Pakistan, a headache for global society.[28]

---

[24] Federal law, "On the federal budget for the year 2001."

[25] Interview with Valeriy Zubov, January 5, 2001.

[26] Ibid.

[27] Kray's special program for years 2000–02, "Creation of semiconductor silicon production in Krasnoyarsk Kray." Available at <http://www.krskstate.ru/dela3/ 18.cfm?nid=200> as of February 15, 2004.

[28] As cited in Nikolay Petrov, *Aleksandr Lebed v Krasnoyarskom kraye* (Moscow: Carnegie Center, 1999). The alleged impetus for this letter, which became fodder for the Russian and international media, was the effort on the part of the wives of the officers and warrant officers to prevent their husbands' from assuming watch duty, in order to protest delays in salary remittances and the allocation of poor-quality food products. This, however, was

As a retired general, Lebed clearly understood that any disruption of the unified command of the SRF was equivalent to a serious breach of Russia's national security. In fact, Lebed did not intend to convey an interest in subordinating the SRF detachment directly to the jurisdiction of the kray. Rather, he seized upon the situation as an opportunity to demonstrate his political mettle at the federal level, while deflecting responsibility within the region. Given the deteriorating economic situation throughout the country at the time, which was raising widespread concerns about the effectiveness of the national leadership, Lebed sincerely believed that the threat to take over an SRF unit, would improve his national stature—or at least keep him in the public eye. Although his gestures caused a public stir, they did not worsen relations between the SRF detachment and local administrations within the closed cities of the kray. According to the SRF division command, relations between the division command and the kray administration simply became more formalized.[29]

On November 17, 1998, Governor Lebed took another shocking step by declaring a moratorium on the import of spent nuclear fuel (SNF) into the kray. Before lifting the moratorium, he demanded that Minatom provide pre-payment for fuel storage at world prices (i.e., increasing payments by a factor of three or four), and that the kray government become a signatory to the contract. Under Lebed's direction, the deputy governor for environmental issues, Aleksandra Kulenkova, prohibited the import of SNF for processing from the Zaporizhzhya nuclear power plant in Ukraine. Kulenkova justified her stance on the basis of unconfirmed reports that the management of GKhK had detected that intermediary structures diverted funds allocated for the storage of Ukrainian fuel and that several State Duma deputies also had been involved in dealings with these nefarious elements. At the same time, deputies of the Zheleznogorsk City Council petitioned the legislative assembly of the kray to lobby the State Duma to change the law "On Environmental Protection" that prohibited the import of foreign SNF for reprocessing. The situation ultimately was resolved by raising the prices for processing Ukrainian SNF.[30]

This moratorium was, to a large extent, aimed at highlighting differences between the Lebed administration's rule and the policies of his predecessor at the regional level. The moratorium, in effect, ran counter to the relationship with central authorities that had been cultivated by the previous kray administration concerning GKhK. Governor Zubov, who headed Krasnoyarsk Kray from 1993 to 1998, favored further development of GKhK. He was convinced that it was necessary to support the Russian nuclear industry, as it was one of the more competitive industries in Russia. In particular, Zubov successfully lobbied the

---

subsequently denied by the division commander. See Aleksey Tarasov, "Opyat babiy bunt," *Izvestiya*, July 25, 1998.

[29] Conversation with the former Deputy Chief of Staff of the SRF Lt. General Vasiliy Lata, Moscow, October 2000.

[30] Interview with a government official in Krasnoyarsk Kray on condition of anonymity, June 2000.

federal government to issue a presidential edict on restructuring and converting GKhK's military production capacity along lines developed by Minatom and the kray administration.[31] During Zubov's tenure as governor, relations between the kray administration and GKhK had been forged on a mutually beneficial basis. Krasnoyarsk Kray received 25 percent of GKhK's income from the processing of SNF. These funds had been allocated towards developing the kray and for the provision of social programs in the area near Zheleznogorsk.

Lebed's subsequent moratorium on the import of SNF from the Zaporizhzhya NPP resonated with local disdain for the previous governor. However, Lebed's position on the issue evolved significantly over a short period of time. Beginning in 1999, when he could no longer expect to obtain a leading role in Moscow but still needed federal support, Lebed began to cooperate increasingly with Minatom. Within the framework of this new orientation, kray administrators, including Lebed, visited the combine several times and supported projects that were advanced by the GKhK management and the leadership of Minatom. By 2001, an agreement was concluded that formalized cooperation between Minatom and the Krasnoyarsk Kray administration.[32] As summed up by GKhK General Director, Vasiliy Zhidkov:

> Since the summer of 1999, cooperation between the combine and legislative and executive agencies of the kray has become certain. There are no problems either with the governor or with the legislative assembly of the kray. In addition, the kray government lobbies for the interests of the combine at the federal level.[33]

In 2001, Lebed openly sided with Minatom as the Federation Council considered bills on the regulation of imports for temporary storage and reprocessing of foreign SNF that again reflected the temporary nature of the 1998 conflict. The kray administration could not ignore the profits involved in plans to store and reprocess SNF in Zheleznogorsk. Former disputes about the environment that arise between GKhK and Krasnoyarsk Kray will be addressed locally, and as such have not become issues of contention with the federal government.[34]

At the same time, relations between the kray administration and the main plant in Zelenogorsk have been stable. This is due largely to the fact that this enterprise has been generally profitable, exporting a large share of its output and intensively expanding its conversion production capacity. The core activity at EKhZ (enrichment and downblending of uranium) also creates less of an environmental

---

[31] Interview with Valeriy Zubov, January 5, 2001. Presidential Edict No. 72, of January 25, 1995, "On state support for structural rebuilding and conversion of the nuclear industry in the city of Zheleznogorsk, Krasnoyarsk Kray."

[32] Interview with Aleksandr Medvedev, director of the legal department of Minatom, January 3, 2001.

[33] Vasiliy Zhidkov, "Poryadok na GKhK sokhranilsya, kak nigde," *Segodnyashnyaya gazeta* (Krasnoyarsk), March 4, 2000.

[34] Interview with a government official in Krasnoyarsk Kray on condition of anonymity, June 2000.

hazard than do activites performed at GKhK. In sum, the relationship between the Krasnoyarsk Kray administration and EKhZ has always been based on the principle of non-interference.

The nuclear policy of Governor Khloponin is motivated by purely economical considerations. It is pertinent to his policy as a whole and it is understandable, since he shifted to public service from the position of a successful former manager of one of the largest Russian enterprises.

Economic considerations also created ground for an anecdotal "nuclear" episode of Khloponin's career. While serving as head of Norilsk Nickel, he seriously considered using mothballed nuclear submarines to deliver cargoes along Russia's northern coastline.[35]

The present Krasnoyarsk Kray administration has also threatened to veto the import of spent nuclear fuel from Ukraine. In 2003, Governor Khloponin asked Ukrainian Prime Minister Victor Yanukovich to pay off Ukraine's $11.7 million debt for storage of spent fuel from Ukrainian NPPs.[36] However, in contrast to Lebed's move, this was not considered to be an act of nuclear separatism, but was welcomed by Minatom, which was eager to collect on this debt.[37] It is not clear how this situation was resolved, but spent fuel deliveries were not interrupted.

In general, closed cities are viewed as a valuable scientific and industrial resource for Krasnoyarsk Kray and enjoy favorable relations with regional authorities.

Finally, the DOE-funded programs aimed at improving the MPC&A systems at both Zheleznogorsk and Zelenogorsk, as well as the NCI program in Zheleznogorsk, have not encountered bureaucratic or other impediments from the Krasnoyarsk Kray administration. According to employees of the Moscow office of the DOE, within the framework of these two programs there is no direct contact with the kray administration, because there is no need for it.[38]

Tomsk Oblast: An Overview

The major industrial sectors in Tomsk Oblast (as a percentage of regional production) include fuel facilities (31.4 percent), chemical and petrochemical plants (25.2 percent), and electrical enterprises (13.9 percent).[39] Oil extraction generates one-third of the budget revenues for the oblast, thus revealing the broad-

[35] "Nikel' I med' uidut pod led" (Nickel and Copper will Go under Ice), Vremya MN. November 23, 1999.

[36] "Gubernator Krasnoyarskogo kraya Rossii Hloponin prosit Yanukovicha pogasit' $11.7 mln. Dolga" (Governor of Krasnoyarsk Kray of Russia Hloponin asks Yanukovich to pay off $11.7 debt), *Ukrainskie Novosti*. August 1, 2003.

[37] "Hloponin primeril lavry Lebedya" (Hloponin Tried On Lebed's Laurels), *Gazeta*. April 7, 2003.

[38] Phone interview with Aleksey Vladimirskiy and Yelena Spitskaya, employees of the Moscow office of the U.S. Department of Energy, October 2000.

[39] Russian State Committee for Statistics, <http://www.gks.ru/regions/statinfo/reg79.asp>.

based structure of the regional economy. The orientation toward raw materials provides the oblast with a relatively high degree of self-sufficiency.[40] In addition, the Tomsk Oblast is one of the most significant scientific and educational centers of Siberia. The science and education complex is regarded as a potential lure for investment by the regional administration. This general structure of the regional economy determines the basic forces that have the greatest influence on the socio-economic situation in the oblast and on decisions made by the Tomsk administration. Representatives of the oblast administration typically single out Russia's second largest oil company, Yukos, and the Siberian Chemical Combine as the most influential corporate players in the region.

With respect to political preferences, the residents of the oblast mirror those of Russia as a whole. For example, in the 2000 presidential elections, Vladimir Putin polled approximately the same in Tomsk Oblast as the national average (Russia—52.9 percent, Tomsk Oblast—52.5 percent). There have been slight differences in parliamentary elections. In the 1999 State Duma elections, for instance, representatives of the right-liberal movements (Union of Right Forces and Yabloko) received slightly more votes, while representatives of centrist and left movements (Yedinstvo, Fatherland-All Russia, and the Communist Party) received fewer votes in the region than the Russian average. In general, however, Tomsk Oblast is on the political periphery of Russia and does not attract much attention from national political parties. On the local level, the greatest influence is exerted by the local industrial elites, who do not maintain close associations with specific political organizations. Typically, the overwhelming majority of deputies in the oblast legislative assembly are leaders of enterprises that are considered large by oblast—and sometimes even by national—standards.[41]

The Nuclear Infrastructure of Tomsk Oblast

The nuclear infrastructure of Tomsk Oblast includes two main facilities: the Siberian Chemical Combine (SKhK) and the Scientific Research Institute of Nuclear Physics at the Tomsk Polytechnic Institute, which has a research reactor. In terms of influence on national security issues and on the situation in the region, the most important facility is clearly SKhK. The chemical combine is one of the largest nuclear centers in the world, incorporating practically all elements of the nuclear fuel cycle. This includes sensitive, nuclear material processing lines, as well as facilities engaged in the production, use, storage, transport, and disposition of nuclear weapons and other radioactive products.

SKhK consists of six factories that are responsible for the combine's major production lines. The first is the Isotope Separation Plant for uranium enrichment. This facility also downblends HEU and converts it into LEU. In addition, the plant also produces stable isotopes. A second facility is the Reactor Plant, which produces weapons-grade plutonium. At present, the ADE-4 and ADE-5 reactors

---

[40] Tomsk Oblast Administration website, <http://www.tomsk.gov.ru/db/web.page?pid=62>.
[41] Interview with Viktor Svinin, a journalist for *Tomskiy Vestnik*, Tomsk, September 2001.

that are scheduled to be taken off line by December 31, 2005 (under a U.S.-Russian agreement), are used to produce heat and electricity for the region. The reactors now provide 30-35 percent of home heating needs for Tomsk and more than 50 percent of the requirements for the city of Seversk and the production facilities of the combine. A third facility is Reactor Plant 5. This plant is slated for renovations, and thus the focus of attention has been on shutting down the reactors for AES-1 and the related production facilities as a whole, as well as organizing non-military production and preparing for the construction of a new AST-500 reactor. SKhK also has a Radiochemical Plant that reprocesses irradiated uranium in order to extract uranium and plutonium. At present, the extracted plutonium dioxide (not purified for weapons use) is housed in a special storage facility. Also located at the facility is the Sublimation Plant, which produces uranium oxide and hexafluoride. The raw materials used at this plant include both natural and regenerated uranium that has been purified at the Radiochemical Plant. Finally, there is the Chemical-Metallurgic Plant that is responsible for the disposition of special products and the processing of extracted HEU. The plant was opened in 1961 for production of metal items from uranium and plutonium. In addition to the six component facilities at SKhK, there are other divisions that support enterprise activities and allow it to be fairly self-sufficient.

In connection with the significant decrease in the last decade of state defense orders, several conversion projects have developed at SKhK that are not related to nuclear materials. That said, most of the revenue of the enterprise is derived from uranium enrichment services and the downblending of HEU within the framework of the U.S.-Russian HEU-LEU agreement. At the same time, the Siberian Chemical Combine is involved in significant foreign commercial activities. SKhK's international partners include such companies as Cogema, URENCO, and BNFL. In 2001, 48 percent of its production was exported.[42]

The nature of SKhK's production and its foreign contacts allow it to maintain a stable position in the civilian market even in the absence of state defense orders. These factors have contributed to stable socio-economic conditions both at SKhK and in Seversk as a whole. Over the last years, the number of personnel has remained more or less constant at 15,500 to 16,000 people. SKhK could lower the average age of the workforce by creating more attractive conditions for young specialists who come to the combine and by providing regular supplementary payments to retiring employees. In addition, it is already becoming possible for SKhK to be selective in hiring specialists, as there is intense competition for employment at the combine.[43] SKhK was a direct beneficiary of the general improvement of the Russian economic situation beginning in 2000 and 2001. In 2001, according to preliminary estimates, the closed city of Seversk collected twice as much in taxes as it did in 2000 (110 million rubles was received in 2000; and

---

[42] Press Release of the Siberian Chemical Combine, <http://www.shk.tsk.ru/press/press.shtml?19092001>, September 19, 2001.

[43] Interview with the dean of the department of technical physics at Tomsk Polytechnic Institute, September 2001.

210 million rubles in 2001). Moreover, investment in the city more than doubled during this period. As noted by SKhK General Director Valeriy Larin, the main problem for the combine now is its limited production capacity.[44]

In addition to commercial contracts, the combine has been involved in a number of joint Russian-U.S. projects in the sphere of nonproliferation of weapons of mass destruction. These projects have involved completion of a system of nuclear MPC&A, work conducted under the HEU-LEU agreement, and conversion of the plutonium production reactors.[45] Also, in the spring of 2003, it was announced that SKhK, instead of GKhK, would be the site for a MOX-fuel production plant to be constructed under U.S.-Russian plutonium disposition agreement. Commencement of construction is scheduled for the first half of 2005.[46] However, this project has been put on hold now due to a lack of funding and the failure of U.S.-Russian officials to resolve a dispute over liability.

The combine also recently kicked off a major public outreach program to market the nuclear sector. This campaign assumed an especially high profile after Larin became general director. It also generally corresponded to Minatom's public relations initiatives. A local effort was conducted via traditional advertisements on the streets of Tomsk, a series of educational events (such as briefings for journalists), and philanthropic activities by the combine to address community problems.

Tomsk Oblast Administration and the Siberian Chemical Combine

The relationship between the Tomsk Oblast administration and SKhK has been forged on the basis of mutual self-interest. SKhK has been primarily interested in working with the regional government to implement conditions favorable to its commercial pursuits, while tapping the influence of the regional administration at the federal level. With respect to the latter, the SKhK leadership has prompted regional authorities to lobby the federal government on behalf of the combine and to woo foreign investment with administrative guarantees that endorse SKhK plans for developing Tomsk Oblast. At the same time, oblast authorities generally have supported the combine for their own pragmatic economic reasons. In the process, however, the oblast leadership has had to balance these economic interests with the "radiophobia" found among the regional population and the need to pay at least lip service to the interests of local environmental groups.

---

[44] Interview with General Director of the Siberian Chemical Combine Valeriy Larin, September 13, 2001.

[45] Aside from its nonproliferation significance, work on the realization of the HEU-LEU agreement is commercial and accounts for a significant portion of the income of the enterprise.

[46] "Stroitel'stvo zavoda po proizvodstvu MOKS-topliva v Rossii nachnetsya v 2005 godu" (Construction of MOX-fuel production plant in Russia will start in 2005), Interfax. May 26, 2003.

The Siberian Chemical Combine actively participates in the political life of the region. In the oblast parliament, Seversk is represented by four deputies, two of whom are employees at the combine, and two others who represent the municipality and internal affairs bureau of Seversk.[47] In general, the Tomsk Oblast Duma has reacted favorably to the demands of the combine. The legislature, in fact, has approved the legal underpinnings for institutionalizing the combine's relations with the oblast. Yet, in the mid-1990s, the local assembly made an abrupt shift, and moved to restrict the local influence of SKhK and Minatom. It passed legislation that instructs the oblast government specifically to:

> forbid Minatom officials to deliver and import to the territory of the Siberian Chemical Combine fissile materials from warheads dismantled according to the normal schedule from other regions of the country and abroad. Should Minatom issue decrees and take action that diverge from this ban, the leadership of the combine must immediately inform federal and oblast authorities about this.[48]

There also was a decision to petition the Russian State Duma with a bill "On the Status of Cities and Settlements Located in the Zone of Influence of Nuclear-Technology Defense Complex Enterprises," which stipulated a reallocation of finances in the nuclear industry. These initiatives revisited attempts that were taken in the early 1990s, when the combine was in a weaker economic and political situation.

One of the most important areas of cooperation for the Tomsk administration and SKhK has been in the development of power engineering. To date, despite significant hydrocarbon reserves on its territory, Tomsk Oblast has experienced an energy deficit and has had to import roughly 60 percent of its energy from neighboring regions. Accordingly, development of power engineering has been one of the priorities of oblast authorities. The oblast administration has hoped to develop non-nuclear energy with the help of the oil and gas companies working in the oblast (Yukos, Vostokgazprom). Yet, in the short-term the oblast has had to remain heavily dependent on SKhK for electrical power. The reactors at SKhK supply electricity and heat to one-third of the oblast's capital—the city of Tomsk. Tomsk residents (who are considered "residents of the SKhK zone of influence") have typically received a discount. Moreover, the combine has taken the initiative to replace an aging heat pipeline connecting Seversk and Tomsk. Recognizing the significance of the pipeline for Tomsk, as well as the need to bolster its public image, the combine leadership has undertaken to carry out this construction.

There has also been much discussion in the oblast regarding the possibility of building an AST-500 nuclear heat plant at SKhK to cover the generating capacity that will be lost after the currently operating reactors are taken out of service by

[47] Despite the fact that these two deputies are not employees of the combine, considering the influence of the combine on the city, it is difficult to imagine that they could win elections in Seversk if they were not completely loyal to the SKhK management.

[48] See decision of October 20, 1994, "On proposed construction of a storage facility for fissile materials from dismantled nuclear warheads."

December 31, 2005. The combine has fulfilled all requirements for building a new nuclear power station. In addition, SKhK was able to sway public opinion in favor of construction of the AST-500. The oblast administration, supporting the aims of SKhK, went even further, as Tomsk Oblast Governor Viktor Kress sent a letter to the Russian government requesting assistance in constructing a nuclear power plant with VVER-1000 reactors. However, the future of nuclear electricity production in the region is unclear. The plutonium production reactors will be replaced with non-nuclear electricity production capacity, but that is the only major energy construction project currently in Tomsk Oblast.

In addition, oblast authorities have joined forces with SKhK to promote the economic diversification of the closed city of Seversk. The most significant non-nuclear development at SKhK is its involvement in projects at the nearby Tomsk Petrochemical Combine. Other civilian activities include the processing of timber at SKhK and a medical equipment manufacturing plant.[49]

It is interesting to note that the preferential tax status of the ZATO went largely unrealized. The introduction of the preferred tax regime did not lead to business development in Seversk, nor did it attract outside businesses.[50] This was due largely to the determination of the oblast to reverse the traditional practice of allowing SKhK to withhold tax payments (totaling 1.5 million rubles) owed to regional coffers. Moreover, a significant number of Seversk residents lived close enough to work in Tomsk, and their residence permits from the closed city enabled them to avoid paying taxes anyway.

Tomsk Oblast authorities regard SKhK as an attractive object for investment. Indeed, Seversk has great potential to become a high technology hub that—together with the significant scientific and educational potential of Tomsk—could serve as a stimulus for transforming the oblast from a "raw materials" region to a developed scientific-industrial center. The oblast administration thus began to implement a progressive program aimed at supporting and legally protecting investment. This has included passage of a local law, "On Innovative Activity in Tomsk Oblast," setting forth the organizational, legal, and economic conditions and guarantees for entrepreneurial activities and regulating relations between the participants in such activities and Tomsk oblast authorities.[51]

In general, the standard of living in Seversk has been higher than in the oblast center and the region as a whole. This is acknowledged widely by the leaders of the oblast and the town, local journalists, and even the residents. This relatively high level of social welfare ultimately has resulted in the transfer of several villages along the borders of the ZATO into the jurisdiction of the municipality of Seversk.

At the same time, the environmental safety of the nuclear facilities at SKhK is a cause of concern for the authorities and residents of Tomsk Oblast. In the early

---

[49] Interview with Deputy Governor of Tomsk Oblast Leonid Bystritskiy, September 17, 2001.

[50] Interview with the General Director of SKhK Valeriy Larin, September 13, 2001.

[51] See Tomsk government website, <http://www.tomsk.gov.ru/db/web.page?pid=2109>.

1990s, environmental groups had a significant influence on public opinion and on regional authorities. Public anti-nuclear sentiments forced SKhK and regional authorities to oppose the construction of a Fissile Material Storage Facility in Seversk in 1992, which was finally moved to Ozersk. Another serious test for the nuclear complex in Seversk was the 1993 explosion at the SKhK Radiochemical Plant. This incident triggered a serious review of SKhK's safety and security practices. To reduce public concerns, SKhK conducted monthly briefings for the press, organized visits to the combine, and adopted a number of measures directed at improving the security infrastructure in case of an accident at the combine.

One of the programs implemented by the Tomsk Oblast administration has been the Tomsk Regional Initiative. This began as a partnership program between the United States and Russia that was intended to encourage and support social programs and the improvement of conditions for private investment. The initiative arose from the eighth meeting of the U.S.-Russian Joint Commission on Economic and Technological Cooperation in February 1997 and was conceived as a broad program of cooperation between the United States and Russia, regional representatives, and the private sectors of both countries. In each of the selected regions, the U.S. State Department, the U.S. Department of Commerce, the U.S. Agency for International Development, and the U.S. Information Agency, along with other agencies, are cooperating with regional representatives to work out a combination of measures for improving the business climate, attracting investment, improving social programs and supporting democracy.[52] This initiative stipulates that, "Along with Tomsk regional authorities, the coordinator of the regional initiative will look for ways to gain positive experience from programs taking place under the Tomsk regional initiative and facilitate activities for the nonproliferation of nuclear weapons." Related documents emphasize that the U.S. Cooperative Threat Reduction Program has had a positive influence on the investment climate in the region.

Tomsk officials also acknowledge that the liberalization of economic relations and development of a democratic society will contribute to the nonproliferation of weapons of mass destruction. The oblast administration has not possessed the legislative authority to facilitate nuclear exports from SKhK or to supervise them. Control over the export activities of SKhK is generally based in Moscow with the mechanisms described in the law "On Export Control" (1999). A branch office of the Ministry for Economic Development and Trade for the Western Siberian region operates in Tomsk. It is authorized expressly to oversee export and import of goods subject to non-tariff regulation and to cooperate with other federal executive agencies in instituting technical measures to regulate the export and import of goods, as well as to manage non-tariff regulation of foreign trade, including the issuance of export-import licenses, and the processing of documents for barter

[52] This also included regional initiatives in Novgorod, Samara, and Sakhalin Oblasts and Khabarovsk Kray.

transactions and other permits.[53] For example, the Tomsk customs service has a division for controlling fissile and radioactive materials.

SKhK and Local Interest Groups

Aside from the Siberian Chemical Combine, the most influential organizations in Tomsk Oblast are the Yukos and Vostokgazprom companies. Because the activities of Yukos and SKhK do not overlap, there have been no serious conflicts of interest to date. Yet, according to former SKhK Director Larin, Yukos may be interested in future cooperation with the combine in order to gain access to SKhK's production and repair facilities, which could service the oil company's equipment. Relations between SKhK and Vostokgazprom, however, have been entirely different, with the two companies embroiled in a protracted dispute over the issue of control of the Tomsk Oil and Chemical Combine (TNKhK).

Nuclear Decentralization in the Siberian Okrug: From Threat to Opportunity

The analysis above suggests two possible models for cooperation between regionally based nuclear facilities and their respective local administrations. Both cases are characterized by the absence of extreme forms of nuclear regionalism (separatism). But the reasons for this outcome differ. In Krasnoyarsk Kray, the primary obstacle lies in the practical inability of the regional administration to commit itself to goal-oriented, long-range actions in the political and economic spheres. Aleksandr Lebed's primary attention as governor of Krasnoyarsk Kray was devoted to fighting against other influential groups in the kray, in which no possibility of compromise was obtainable. His nuclear bombast was highly propagandist and was employed mainly to boost his campaign to succeed Yeltsin as president of Russia. The relatively low level of trust in General Lebed exhibited in Krasnoyarsk Kray and his unflattering image at the federal level undermined the credibility of any separatist claims. As his tenure wore on, Lebed effectively depleted his political or economic resources for aggressive action. In addition, as governor of Krasnoyarsk Kray, he was beholden to support unconditionally the initiatives of the leadership of GKhK, in order to increase his chances for support from the federal center in his struggle against his regional opponents. The kray administration had to have working, if not always good, relations with GKhK and EKhZ, since both facilities are located within the kray and are of great importance to Russia's national security. Eventually, relations between the regional administration and "nuclear" actors in the region entered into a "pragmatic" phase with the election of Khloponin as governor.

The situation in Tomsk Oblast has been significantly different. There, the local elite is comprised of the leaders of those industrial enterprises that have the most influence in the region. This allows SKhK and the oblast administration to base

---

53 See Tomsk government website, <http://www.tomsk.gov.ru/db/web.page?pid=2109>.

their relations primarily on mutual economic interests, which happen to be largely compatible with those of the federal center. In Tomsk, there is a system of "understanding" aimed at individual gain: the oblast administration strengthens the socio-economic conditions and develops the region, while SKhK establishes a favorable environment for its activities and receives support from a governor who is fairly influential on the federal level.

The essential difference in the political influence of GKhK and SKhK in their respective regions can be explained mainly by the ratio of the population of the closed cities to that of the region. In Krasnoyarsk Kray, the ratio is much lower than in Tomsk Oblast, as is the local population's dependence on the nuclear facilities. SKhK contributes significantly more to the daily lives of the residents of Tomsk than GKhK does for the people and government of Krasnoyarsk Kray. At the same time, because all of the nuclear establishments in both regions (SKhK, GKhK, and EKhZ) are state unitary enterprises and are located on the territories of ZATOs, they all enjoy special tax privileges. The legislation in this area and strong economic ties to the center basically deprive respective regional administrations of any serious economic levers to influence the nuclear enterprises. The Tomsk Oblast administration's hope for changing the tax status of Seversk is only a desire, not a demand, which prevents any serious conflicts with the center.

Furthermore, the integration of the enterprises into the structure of Minatom excludes the possibility of their management acquiring greater independence that could otherwise hamper national security and raise risks of the proliferation of nuclear materials and technology. The lack of significant progress in restructuring the system of Minatom nuclear fuel cycle enterprises indicates that, in the near future, management at such enterprises will not undergo significant changes, nor will the nuclear facilities obtain a significant amount of economic freedom from the center.[54]

Because nuclear-related enterprises in the Siberian Federal Okrug are also manufacturing centers, there could be a silver lining. As such, they maintain stable and relatively favorable socio-economic conditions in the ZATO. To a large degree this is related to enterprises, such as SKhK, EKhZ, NZKhK and AEKhK, which retain manufacturing capacities that could be employed successfully in commercial projects. This is less applicable to GKhK, which is oriented primarily toward weapons-related activities.

In addition, the "city forming" status of the enterprises, as well as the closed status of the cities themselves, renders the situation in the ZATOs very dependent on the activities of these enterprises. Any disruption of their work could cause a social crisis. Therefore, efforts to diversify the economies of closed cities are necessary and need to be pursued.

Thus far, nuclear enterprises in the Siberian Federal Okrug have not attracted serious attention from criminal groups. This is mainly a problem for export-

[54] Moreover, the enterprise management often thinks that radical restructuring is unnecessary. For example, SKhK General Director Valeriy Larin likes to refer to himself as a "man of the state."

oriented mining and processing enterprises. This is especially true in Krasnoyarsk Kray, where the level of criminalization of political and economic life has increased. However, it is not possible to rule out completely the future illegal use of funds at nuclear enterprises.

In conclusion, it should be noted that the Siberian Federal Okrug has significant potential to lobbying for the interests of Minatom on the federal level. This should benefit both Minatom, which seeks to establish favorable legislation and public opinion for the work of its enterprises, as well as the constituencies of the Siberian Federal Okrug, which hope to use the scientific and technical potential of the nuclear sphere as a catalyst for their economies. Constructive cooperation among the leadership of these Minatom enterprises, the regional administrations, and federal authorities could improve the socio-economic situation at the local level. That said, it is premature to dismiss the possibility that an imbalance among the interests of the enterprises, the regional governments, and the country as a whole might in the future undermine national security and threaten the interests of the population living near the nuclear facilities.

PART III
THE EXPERIENCE OF U.S. ASSISTANCE PROGRAMS AND CONCLUSIONS

Chapter 10

The Implementation of U.S. Nonproliferation Assistance Programs in Russia's Regions

Michael Jasinski and Charles Thornton

With the imminent break-up of the Soviet Union and in the context of the late-Soviet economic crisis, the U.S. government undertook an assistance program designed to eliminate or reduce threats to the security of weapons of mass destruction (WMD) in the newly independent states (NIS). After the enactment in 1991 of the Soviet Nuclear Threat Reduction Act, the U.S. and Russian governments and their agencies involved in implementing U.S. assistance projects signed a series of agreements to establish a legal framework for assistance activities. These agreements provide a comprehensive set of rights, exemptions, and protections for U.S. assistance personnel and program activities. Agreements were also signed with other NIS countries possessing nuclear weapons with the intention of helping them transfer these WMD to Russia and eliminate their remaining WMD infrastructures.

The two main U.S. agencies that have implemented assistance programs under these agreements in Russia have been the Department of Defense (DOD), which implements the Cooperative Threat Reduction (CTR) program, and the Department of Energy (DOE), which is responsible for such programs as the Nuclear Cities Initiative (NCI), Initiatives for Proliferation Prevention (IPP), and the Materials Protection, Control, and Accounting (MPC&A) program. Other activities are conducted by the U.S. Department of Commerce, the State Department, and the Environmental Protection Agency.

These intergovernmental agreements, however, only established the broad framework for implementing the assistance projects. The actual implementation saw the U.S. agencies and contractors encounter a variety of regional and local factors that sometimes hindered the projects, although on other occasions *facilitated* them and helped overcome barriers at the federal level. The character and importance of these factors varied widely depending on the nature of the project, the relationship between U.S. and federal Russian agencies, and the political situation in the region. Faced with these factors, the U.S. entities have adopted a variety of approaches to either minimize their impact or use them to advance project objectives.

These approaches have varied with the nature of the projects and the relationship between the U.S. agencies and their Russian counterparts. In general, DOD projects have had relatively well-defined and quantifiable objectives, such as the elimination of strategic weapons systems. In contrast, DOE projects, with the exception of the MPC&A program, have pursued less-well-definable objectives, which sometimes require structural changes (such as downsizing or conversion) in the Russian enterprises. These differences also extend to the patterns of interaction between U.S. and Russian federal government agencies. DOD's Russian partners include not only the Ministry of Defense (MOD), but also the Ministry of Atomic Energy (Minatom), the Russian Aerospace Agency (RASA), and others. DOE, on the other hand, deals almost exclusively with Minatom (again, with the exception of the MPC&A program, where its partner is the MOD). Minatom's virtual monopoly of power in its nuclear closed cities, where the NCI and IPP projects are implemented, reduces the problem of interagency cooperation. Due to these differences, CTR project managers and U.S. contractors have tended to become closely involved in managing the regional issues, in the absence of a coordinating Russian government agency. DOE instead has elected to rely on Minatom to overcome regional obstacles. As a result, DOE program officials have been less exposed to regional issues, although their programs have also been affected.

Another lesson of the U.S. experience with regional factors has been that regional factors have not been always an obstacle to project implementation. In many instances, regional and local government entities have helped remove barriers put in place by central authorities. Russian subcontractors have become a constituency interested in continuing these programs, while local environmental groups have sought to draw attention to such urgent issues as nuclear submarine dismantlement.

This chapter first provides some critical background information regarding decentralization of Russia's nuclear complex and its relevance to both DOD and DOE programs. It then analyzes the impact of regional factors specifically on the DOD's CTR program, by evaluating the influence of regional and local governmental and nongovernmental entities on project implementation and assessing how DOD has adapted its programs to deal with these issues. Finally, the chapter examines DOE's experience in Russia's regions and how has it differed from that of DOD. The experiences of the State Department and other lesser U.S. programs in implementing nonproliferation assistance projects are not considered here, due to limits of space and a desire to focus on the major trends in U.S. assistance programs.

Background Issues

In spite of the agreements reached between the various U.S. and Russian government organizations, U.S. officials and contractors responsible for project implementation found these agreements were insufficient to guarantee successful implementation. As a result of Russia's evolving political and economic landscape,

U.S. program managers and contractors were compelled to deal with local agencies that had authority over such key issues as site access, environmental compliance, and the hiring of local contractors. Patterns of interaction between U.S. entities, Russian federal authorities, and regional and local actors varied greatly from one project to the next, and even from one region to the next, presenting additional complications. As a result of this wide range of factors affecting U.S. assistance projects, the government-to-government and agency-to-agency agreements could not account for all variations inherent in the relations between the central government and regional entities. While in some instances the regional and local influences have added complexity, and delayed implementation, the influence of regional factors has not always been a negative one. In many instances, regional and local authorities have proven willing to go beyond the constraints imposed by central authorities and have even lobbied the center on behalf of some projects.

Regional influences on project implementation were the result of the relatively high degree of autonomy that Russia's regions achieved during the rather chaotic decentralization process that characterized the Yeltsin era. In the 1990s, the Russian federal government concluded separate agreements with individual Russian Federation components on taxation and other financial issues, and permitted the regions considerable legislative leeway as well. While under President Vladimir Putin the Russian government has made a concerted effort to cancel these agreements and bring each region's constitution in line with the constitution of the Russian Federation, this effort has not been wholly effective, as regional governors have defended their prerogatives and sought to retain control over tax revenues. The contribution each region makes to the federal treasury also affects the degree of control the federal government has over that region. As a rule, regions that are net contributors of tax revenues to the center enjoy a higher degree of autonomy than regions that are net recipients of tax distributions. Net donors of revenues are more likely to be able to influence decisions being made that affect program implementation.[1]

The central government has therefore lost much of its ability to reward and punish the provinces, due to the greater political autonomy gained by regional governors, the reduced financial clout of the center, and other considerations. Many responsibilities of the central government have devolved to the local bodies and became additional sources of local power.[2]

Another aspect of Russia's political changes is the fact that the agencies of the Russian federal government are not as centralized as they were during the Soviet era. Rather than presenting a unified front when dealing with regional authorities, they sometimes work at cross-purposes. Although President Putin has had a measure of success in consolidating the federal-level agencies, especially the law enforcement bodies, the problem persists today. Moreover, federal agencies have often not worked out approaches for dealing with problems besetting the nuclear or

---

[1] Interview with congressional staffer, Washington, D.C., August 2001.

[2] Interview with U.S. Department of Defense CTR technical program manager, Washington, D.C., August 2001.

defense establishments, thus undermining their credibility. Regional and local leaders are sometimes able to exploit inter-agency differences by relying on their friends in federal agencies to countermand orders not to their liking. The weakening of the Soviet-era vertical law enforcement system is part of that phenomenon.[3]

The Russian government's ability to deal with regional concerns is further weakened by the reorganizations of government agencies tasked with project implementation. For example, the CTR program's Strategic Offensive Arms Elimination (SOAE) project was initially the responsibility of the Russian Committee for the Defense Industry, which later became the Ministry of the Defense Industry. This ministry was eliminated in 1997 and the responsibility transferred to the Ministry of the Economy, while its implementation was entrusted to the Russian Aerospace Agency (RASA). Such rapid organizational change has impeded continuity.

Even the military could not escape the consequences of the growth of regional influence. In the 1990s, the Russian government found itself unable to finance the military establishment it inherited from the USSR, resulting in a greater role for regional authorities in supporting the military. Due to the military's general impoverishment, and in the absence of clear rules governing the disposal of military property, many officers entered into business dealings (of varying degrees of legality) with local politicians or businessmen to improve the funding situation for their units or, in some cases, themselves. In return for selling trucks, fuel, and other equipment, military units obtained funds they needed. In spite of the centralized nature of the military, inadequate finances frequently forced isolated garrisons to fend for themselves, and their commanders to enter into deals with local political or business entities to secure basic necessities. As a result the local entities acquired a means of influencing military units, and military leaders a vested interest in the fortunes of commercial entities. These activities have created regional business networks in which U.S. agencies implementing assistance programs (whether DOD or DOE) have had to operate.[4] Interaction between military officers and local leaders has also varied depending on the priorities of the officers in question. In some cases, officers enter into deals with local governments or businesses out of necessity—in order to provide basic services for their troops and their families. In other cases, they do so for personal gain.

---

[3] Interview with Congressional staff member, Washington, D.C., August 2001.

[4] Interview with U.S. Department of Defense CTR technical program manager, Washington, D.C., August 2001.

The CTR Program

Local Authorities and Assistance Project Implementation

The diminished ability of the Russian government to impose its will on the regions has forced the U.S. project managers to play an active role in dealing with regional authorities. It burdened the managers with increased decisionmaking responsibilities. U.S. project managers and contractors found themselves having to coordinate the activities of the various agencies of the Russian government (including their regional offices, which sometimes were staffed with personnel close to the regional authorities), and between federal agencies and local authorities. As a result, managers have to deal with authorities in Moscow, base commanders, construction teams, regional political leaders, and other officials. The problem of coordination is made more difficult by the large number of senior Russian officials or officers (at both the regional and federal levels) responsible for specific issues related to project implementation. The situation has diffused responsibilities that exist in many regions, making it difficult at times to ascertain which of the officials have been empowered to perform the functions of regional "middle men," and which have been trying to throw up obstacles to implementation for personal gain.[5]

As discussed earlier, CTR project exposure to regional influences has varied from project to project. For example, regional influences have been fairly limited in Warhead Protection, Control, and Accountability (WPC&A) projects conducted to assist the 12th Main Directorate, a closely-knit and centralized organization with an established network of contractors. There has been equally little potential for regional involvement in the nuclear warhead transportation safety upgrade projects, as contracts have been awarded to Russian firms that have experience and certifications for required equipment, and their products have been used throughout the 12th Main Directorate. Regional influences have been limited to selecting third parties to perform verification and forwarding agents who monitor train shipments.

Other CTR projects, of which the Inter-Continental Ballistic Missile (ICBM) silo demolition project is representative, have had greater exposure to regional influences. U.S. contractors have had to rely on local subcontractors with clearances to work on the missile bases, and there have also been environmental issues that have attracted the attention of the general public and the regional and local governments. Moreover, ICBM elimination projects are usually terminated by turning over the former Strategic Rocket Forces (SRF) bases to local authorities, which are entitled to demand certain ecological standards be met by the contractors.

The friction between central and regional authorities has been at times used as an excuse for failing to meet project obligations. Minatom, for example, has blamed regional Federal Security Service (FSB) officers for failing to issue necessary clearances related to the construction of the Mayak Fissile Material

---

[5] Ibid.

Storage Facility (FMSF). It is not clear whether such problems reflect actual conflict between regional and federal offices, or are the symptoms of inter-agency struggles.[6] These delays may be caused at the behest of Minatom itself, if it wants to delay the project for some reason. In other cases, the central and local authorities issue contradicting orders, with damaging consequences for the project. In one case where a local government issued an access permit, shortly afterwards, central authorities countermanded the order.[7] These and other incidents have shown the need to build trust between U.S. officials and contractors and local agencies. When local governments are instrumental to issuing permits and delays drive up project costs, establishing trust early in the operation can prevent cost escalation.[8]

On the other hand, there have been numerous cases of regional and local authorities assisting in project implementation. Such is the case in the Primorskiy Kray, where the office of the Pacific Fleet commander-in-chief has had considerable political influence and generally good relations with both Moscow officials and U.S. representatives.[9] Moreover, the local shipyards have been active in lobbying for submarine dismantlement work. In other cases, regional governments have taken measures to facilitate implementation without waiting for instructions from Moscow. The maverick former Krasnoyarsk Kray Governor Aleksandr Lebed even went so far as to claim that his statements on the dire situation on SRF bases in his region helped the U.S. government secure additional CTR funds from the U.S. Congress. Regional assistance is in many cases motivated by the regional authorities' desire to attract lucrative CTR projects, in others by a desire to demonstrate political independence from Moscow.[10] The local authorities' willingness to bend the rules for the sake of CTR project implementation also depends on the central government's willingness to tolerate infractions. However, after 2000, that willingness has been on the decrease due to President Putin's attempts to bring the regions under centralized control. As a result, there has been greater deference to federal authorities.[11]

Central government officials in general appear to be aware of regional problems and, when unable to influence the situation, have sometimes acted as a bridge between local officials and U.S. participants.[12] With time, the awareness by central government agencies of regional constraints and their ability to operate within them has improved. In some cases ministry officials have managed

---

[6] Interview with U.S. Department of Defense Mayak FMSF technical program manager, Washington, D.C., August 2001.

[7] Interview with Rose Gottemoeller, Washington, D.C., September 2001.

[8] Interview with senior U.S. Department of Defense official, Washington, D.C., August 2001.

[9] Interview with Rose Gottemoeller, Washington, D.C., September 2001.

[10] Interview with senior U.S. Department of Defense official, Washington, D.C., August 2001.

[11] Interview with Rose Gottemoeller, Washington, D.C., September 2001.

[12] Interview with senior U.S. Department of Defense official, Washington, D.C., August 2001.

successfully to become part of the decisionmaking process, acting as intermediaries among the U.S. government, contractor representatives, and local authorities.[13]

Site Access Issues

The implementation of CTR projects in the Russian Federation was made more difficult by the fact that, in contrast with Ukraine, Belarus, and Kazakhstan, which gave up their nuclear forces, Russia still had secrets to protect, thus complicating issues associated with site access. Regional influences manifested themselves in a twofold manner. While, in some cases, the reluctance of central authorities (usually the MOD or Minatom) to allow U.S. personnel to visit sites necessitated the selection of local subcontractors with clearances, in other projects it was the local authorities that posed site access difficulties or, alternatively, facilitated access.

Thus, for example, while Russian bomber elimination, which did not reveal operational secrets, could be performed by U.S. contractors, ICBM silo eliminations proved to be a more complex issue. Due to the Russian government's unwillingness to clear U.S. contractors, the U.S. firms had to select local firms to perform the work. While the closed nature of the sites reduced the organized crime threat, the need for clearances reduced the pool of qualified subcontractor candidates, in some cases to a very small number. The reduction in the number of candidates made negotiations more difficult, especially in cases where only one local contractor had the clearance to work at a given site.[14]

Since the MOD controls access to its facilities, it can sometimes use this right as a means of influencing the decisions made by U.S. officials. During the subcontractor selection, the MOD might exclude firms for such reasons as inability to obtain a clearance, criminal background, or lack of expertise. There have been cases reported of the MOD blocking facility access in order to ensure that a favored company was hired. Exercise of this authority has given rise to allegations of collusion between MOD officials and firms. When the responsibility for silo eliminations was transferred to RASA, however, it acquired the responsibility for clearing subcontractors.[15]

One variation on the theme of local subcontractors has been the proposed use of "facilitating agents" (firms possessing 12th Main Directorate clearances) to provide security upgrades to nuclear weapons storage facilities without allowing Defense Threat Reduction Agency (DTRA) personnel permission to visit the sites to verify the work. The verification would be performed through photographs provided by the "facilitating agents." The MOD granted access to only one facility, at Aleysk (the location of an SS-18 division being disbanded) for DTRA personnel to verify the work. This initiative was proposed by 12th Main Directorate Chief

---

[13] Ibid.

[14] Ibid.

[15] Interview with U.S. Department of Defense CTR Technical Program Manager, Washington, D.C., August 2001.

General Igor Valynkin, who was anxious to provide security upgrades to specific facilities deemed to be threatened.[16]

In cases where the CTR project is to be implemented not on a military base but on a lot belonging to the local authorities, site access issues can also arise. In the case of the proposed Votkinsk solid rocket fuel incinerator, local authorities pressured Russian federal agencies to provide payments for site access, even though the site was a state-owned facility.[17]

Projects that take place in the so-called "closed cities" can create additional difficulties by adding another layer of required clearances. Although mayors of closed cities are authorized to grant 24-hour access to the city, this authority does not extend to granting permits to enter specific facilities.[18] While closed cities proved to be more sensitive to Moscow's preferences, since they are still more dependent on central funding, in some cases even closed cities, particularly those that have diversified their industries and depend on various business enterprises for revenue, have shown greater independence from the central government.[19] In spite of the limitations inherent in their status, occasionally these concerns have been outweighed by the perceived benefits and prestige associated with it. In some cases, the closed cities have opted for this status because it freed them of having to pay certain types of taxes.[20] Although these privileges have been revoked, retaining the closed city status may be a way to ensure that financial profits made from the scarce resources there (fissile materials, etc.) stay in the city and are not distributed within the oblast.[21]

For example, on May 24, 2000, the Arkhangelsk Oblast press office reported that Governor Anatoliy Yefremov had requested the federal government to give Severodvinsk, the home of facilities involved in SSBN dismantlement projects, closed city status. If granted this status, Severodvinsk would have come under federal jurisdiction and would have received funding from the federal budget and tax exemptions. Supporters of this initiative estimated that the city's budget would double from 300 million rubles to 600 million rubles, and the city's taxes would decrease by 41 million rubles. Oblast authorities expressed support for the initiative, since it would relieve them of responsibility for the defense enterprises

---

[16] Bill Moon, "Nuclear Weapon Storage Site (NWS&S) Enhancements Prior to Agreed Procedures For Site Access to Russian NWS&S," Memorandum for the Office of the Assistant Secretary of Defense International Security Policy Threat Reduction Policy, August 2001.

[17] Interview with U.S. Department of Defense SOAE technical program manager, Washington, D.C., August 2001.

[18] Interview with Rose Gottemoeller, Washington, D.C., September 2001.

[19] Interview with senior U.S. Department of Defense official, Washington, D.C., August 2001.

[20] Ibid.

[21] Interview with U.S. Department of Defense Mayak FMSF technical program manager, Washington, D.C., August 2001.

in Severodvinsk.[22] Although closed city status is usually viewed as an obstacle to outside investment, in the case of Severodvinsk, some had hoped that closed city status would improve the business climate in Severodvinsk and attract foreign investment.[23] This proposal was never realized, however.

Local Contractor Participation in CTR Projects

One of the initial problems that hampered U.S. assistance programs was the "buy American" provision in the CTR legislation, which required the U.S. government to use the goods and services of only U.S. companies, unless there was no alternative. However, the Russian side expected financial benefits in addition to security. Virtually excluding Russian subcontractors from the process eliminated them as a possible constituency in Russian regions for lobbying the local and/or central government to facilitate program implementation.[24] Such rules have now been relaxed.

Under the terms of the CTR program, private firms in the NIS can be involved in project implementation as subcontractors only on the basis of fixed price contracts. U.S. contractors usually negotiate prices for services in advance, and reimburse the subcontractors only for legitimate expenses. Disbursing funds to the U.S. integrating contractor, rather than directly to Russian subcontracting firms, increases the bargaining power of the U.S. side.[25]

Given that the potential profits for enterprises involved in CTR project implementation can be quite large, competition for such contracts can become intense. Facilities involved in CTR project implementation frequently receive scant government funding. The level of state defense orders has dropped sharply during the 1990s, and, even when such orders are placed, the Russian government often fails to pay for them, resulting in indebtedness.[26] In one instance, a case of potential subcontractors sabotaging each other's attempts to be included in the project was reported. This took place in Perm, where an incinerator for solid rocket fuel extracted from ICBMs was to be established. When the project was moved by the Russian government to Votkinsk, U.S. government officials met with the local mayor to ensure this would not happen again.[27] There have also been reports that

---

[22] "Severodvinsk mozhet obresti status zakrytogo territorialnogo obrazovaniya," Regions.ru website, May 24, 2000; in National News Service website, <http://nel.nns.ru>.

[23] "Severodvinsk to become closed city," Bellona Foundation website, May 24, 2000, <http://www.bellona.no>.

[24] Kenley Butler, "Russia: Government-to-Government Program," Nuclear Threat Initiative website, April 13, 2001, <http://www.nti.org/db/nisprofs/russia/forasst/doe/govtogov.htm>.

[25] Interview with U.S. Department of Defense CTR Technical Program Manager, Washington, D.C., August 2001.

[26] Interview with senior U.S. Department of Defense official, Washington, D.C., August 2001.

[27] Ibid.

cash-strapped subcontractors sometimes attempt to stretch out project schedules in order to maximize the profits.[28]

In these conditions, the local elites become interested in influencing the course of CTR projects, particularly ones that involve Russian subcontractors since they offer a means of exercising patronage. There have been allegations that officials involved in subcontractor selection have received kickbacks from favored firms. In other cases, local military commanders have insisted on using firms on whose board of directors they serve. Local officials have at times sought to limit the number of firms in the selection process to eliminate the possibility of contracts being awarded to non-favored firms. In addition, local governments have sought to tax the assistance projects. In May 1999, the Russian State Duma passed a law exempting all nonproliferation assistance programs from taxes, in part thanks to the efforts of U.S. negotiators. However, in view of the difficulties the Russian government has in ensuring its laws are implemented in the regions, it is not certain that the law will have the desired effect. While direct assistance was not subject to taxes, contractors and Russian subcontractors were. Salaries paid by contractors were also taxable, and the contractors were obligated to pay payroll taxes.[29]

The central government's approach to specific projects also depends on the level of priority attached to the outcome. In the case of the Mayak FMSF, its completion is reportedly not considered a top priority for Minatom, since the facility is not going to be a source of income for the ministry.[30] Minatom's lack of interest has been identified as one of the reasons for the difficulties Mayak construction encountered. In this instance, the subcontractors, who had a direct interest in the success of the project, proved to be more enthusiastic than central authorities. The Sarov-based All-Russian Scientific Research Institute of Experimental Physics (VNIIEF) was able to go further in joint cooperation efforts than Minatom was willing to tolerate, including in such ventures as theoretical transparency studies for the FMSF. Faced with Minatom intransigence, VNIIEF has been suggesting alternative approaches to overcome objections.[31]

Contacts between program managers and local contractors have been used as a venue for ensuring project accountability.[32] Oversight of Russian subcontractors by U.S. contractors brought dividends by assuring that the projects were being carried out in a manner that limited the risk of proliferation. The reported cases of ballistic missile gyroscopes finding their way to Iraq, although not traced to CTR-funded

---

[28] Interview with U.S. Department of Defense Mayak FMSF Technical Program Manager, Washington, D.C., August 2001.

[29] Interview with U.S. Department of Defense CTR technical program manager, Washington, D.C., August 2001.

[30] Interview with U.S. Department of Defense Mayak FMSF technical program manager, Washington, D.C., August 2001.

[31] Ibid.

[32] *Cooperative Threat Reduction Annual Report to Congress*, U.S. Department of Defense, 2001, pp. III–3.

contractors, prompted U.S. contractors to require strict accountability for sensitive devices.[33] Furthermore, Russian enterprises that participate in U.S.-funded projects frequently adopt Western accounting techniques, which allows them to reduce their overhead costs and become more competitive. For example, the use of a fixed-price contract for the submarine elimination project is estimated to have saved about 40–60 percent of the cost by eliminating the general contractor and associated overhead.[34]

Regional Environmental Issues

A number of U.S. assistance projects have encountered resistance on the part of local governmental and even nongovernmental organizations based on environmental considerations. Since the authority for environmental problems devolved to regional governments during the Yeltsin administration, the central government has been compelled to secure permissions for land use and facility construction from regional governments. Given the powerful legal means of influencing the central government that locally formulated ecological requirements represent, these issues sometimes are used as a way of pressuring the central government for further concessions to local interests that could not be obtained through other means.[35] Nevertheless, Russian environmentalists have also been at the forefront of efforts to implement some assistance projects, including the work to eliminate Russian nuclear-powered submarines.

In the solid fuel missile incinerator project, for which the U.S. government issued a tender in 1996 to eliminate over 900 solid-fuel SS-24, SS-25, and SS-N-20 ICBMs and SLBMs, 17,500 metric tons of solid rocket fuel turned out to be affected by environmental objections. Initially, the incinerator was to have been constructed in Perm, where ICBMs were formerly produced. Following intense protests by the oblast's population and leaders, however, the project was transferred to Votkinsk, the site of a plant that specializes in production of a range of solid-fuel missiles. However, the project also ran into considerable opposition in Votkinsk, where ecologists cited U.S. studies claiming the negative effects of rocket fuel incineration on the state of Nevada. In 1999, the city of Votkinsk conducted a referendum on the incinerator, and over 90 percent of participants voted against it. Even though the referendum was later invalidated by the Udmurt election district and the Supreme Court of Udmurtia on the grounds that the referendum had to involve all the residents of Votkinsk Rayon, public resistance to the project did not abate, and the government of the adjacent Perm Oblast also

[33] Interview with U.S. Department of Defense SOAE technical program manager, Washington, D.C., August 2001.

[34] John Lepingwell, Nikolai Sokov, "Strategic Offensive Arms Elimination and Weapons Protection, Control, and Accounting," *The Nonproliferation Review* 7 (Spring 2000), pp. 60–75.

[35] Interview with senior U.S. Department of Defense official, Washington, D.C., August 2001.

voiced its concerns. Ultimately, local resistance to the project in Votkinsk proved too strong to overcome. As a result, in October 2003, RASA announced that the project would be moved back to Perm and that the elimination of solid-fuel missiles would only start in late 2003.[36]

The heptyl liquid rocket fuel conversion project has suffered from similar locally based environmental resistance. Initially planned for the city of Sergiyev Posad, it had to be relocated to Krasnoyarsk as a result of intervention by the Russian Orthodox Church at the behest of local inhabitants. This development has been attributed to the absence of local public relations outreach by the CTR program.[37] The Zlatoust Machine-Building Plant in Miass (Chelyabinsk Oblast), which is involved in liquid-fuel SLBM elimination, has also encountered local ecological resistance. Although the Makeyev State Missile Center certified the safety of elimination processes used at Zlatoust, Miass City Council deputies disagreed, citing the possibility of an accident involving highly toxic liquid rocket fuel components.[38]

Since ICBM silo elimination projects are usually terminated with the transfer of the former missile base's territory to local authorities, these authorities have a vested interest in ensuring elimination work meets their environmental requirements. During the elimination of SS-18 missile silos belonging to the former Aleysk missile division (Altay Kray), the local government closely followed the elimination process and demanded that after the elimination of each missile regiment, the terrain be fully recultivated before work on eliminating additional silos could proceed. The Ministry of Defense and RASA were compelled to prepare a recultivation schedule for these sites.[39] In May 2001, SRF Commander Nikolay Solovtsov signed a protocol on the transfer of the military base to the city. Under the terms of the protocol the SRF was obliged to pay the debt owed by the unit to the city's water works utility. RASA, in turn, was obligated to eliminate all metal structures on the site, recultivate the sites of eliminated silos, and conduct an inspection.[40] Local authorities were justifiably concerned about the methods used to eliminate the silos (various reports described the process as involving the detonation of 3,000 anti-tank mines, or three tons of high explosives per silo), which left the area surrounding the silo unsuitable for

[36] Dina Pyanykh, "Ekologicheski bezopasnyy kompleks unichtozheniya strategicheskikh raket do kontsa goda zarabotayet v Permi," ITAR-TASS, October 17, 2003; in Integrum Techno, <http://www.integrum.ru>.

[37] Interview with senior U.S. Department of Defense official, Washington, D.C., August 2001.

[38] Yevgeniy Tkachenko, "Utilizatsiya rossiyskikh ballisticheskikh raket podvodnykh lodok na Zlatoustovskom mashinostroitelnym zavode priostanovlena na neopredelennyy srok," ITAR-TASS, November 20, 1999; in Integrum Techno, <http://www.integrum.ru/>.

[39] "Podpisan protokol o reformirovanii Aleyskoy raketnoy divizii," Interfax, May 23, 2001.

[40] Barnaul Informatsionnoye Agentstvo, May 23, 2001; in "RVSN's Solovtsov signs protocol transfering property of Aleysk missile division to local and regional civilian authorities," FBIS Document CEP20010523000401.

agricultural use.[41] In addition to state ecological institutions, the demolitions were also overseen by the Altay Kray administration.[42]

The Aleysk missile division elimination was of considerable importance to local authorities, since the SRF unit was an important factor in the city's economy and the missile division's service members and their families accounted for half of its population. Its disbandment would therefore create hardships for the city and, as in the case of many other Russian towns that exist mainly to support a nearby military facility, possibly its disappearance as well. The political leadership of Altay Kray, however, was reportedly keenly interested in the fate of the city and took measures to minimize the impact of the division's disbandment. The kray political leadership negotiated the transfer of the base's housing and infrastructure from the Ministry of Defense to the city, and decided to house the kray's Interior Ministry training center there. The kray government also required that the silo sites be recultivated following their elimination.[43]

Not surprisingly, in view of the hazards inherent in nuclear submarine elimination, these projects are also subject to environmental opposition. In February 2001, the Arkhangelsk Oblast Administration Environmental Protection and Nature Management Board declared the nuclear submarine elimination activities at Severodvinsk's shipyards ecologically unsatisfactory.[44] In July 2001, the board had prohibited the military from towing additional decommissioned nuclear submarines to Severodvinsk because there were already 15 such vessels awaiting elimination in the city's ports.[45] Nevertheless, Russian environmentalist organizations are, as a rule, not so much opposed to the idea of submarine elimination *per se*, but are rather concerned with the details of the implementation. In fact, environmentalists on the Kamchatka Peninsula have been seeking foreign assistance for dismantling deactivated nuclear-powered general-purpose submarines, whose state of neglect poses an ecological threat.

Environmental issues have also arisen in the construction of the Mayak FMSF. The facility was originally planned to be constructed in Seversk. Due to its rejection by Seversk authorities, the project was moved to Chelyabinsk Oblast. In March 1998, the Russian State Committee on the Environment conducted a review of the facility and requested that Mayak and the storage facility contractor submit a

---

41 "Na Altaye zavershena podgotovka k unichtozheniyu shesti shakhtnykh puskovykh ustanovok, osnashchennykh ballisticheskimi raketami," ITAR-TASS, April 9, 2001; in Integrum techno, <http://www.integrum.ru>.

42 Valentin Pavlov, ITAR-TASS, November 2, 2000; in "Russian missile forces start demolishing ICBM missile silos," FBIS Document CEP20001102000441.

43 Center TV, May 31, 2001; in "Russia: Military, civilian authorities discuss future of Siberian town," FBIS Document CEP20010531000073.

44 "Rabota po utilizatsii atomnykh podvodnykh lodok, provodimaya na predpriyatiyakh Severodvinska, priznana neudovletvoritelnoy," *Rosbalt*, February 1, 2001; in MA Foris, February 3, 2001; in Integrum Techno, <http://www.integrum.ru>.

45 Viktor Filippov, "Ustalyye podlodki," *Izvestiya Peterburg*, No. 120, July 7, 2001; in Integrum Techno, <http://www.integrum.ru>.

revised environmental impact statement. Officials at Mayak submitted the requested documents by the end of that March, thereby avoiding interruption of the construction.[46]

Regional laws sometimes also may impede the transportation of radioactive materials. Some regions have already enacted legislation prohibiting transit of nuclear materials through their territories, in violation of the "On Nuclear Energy" federal law. In some cases, the secrecy of the transports has been compromised by the Ministry of Railways' requests for pre-payment of fees for such services.[47]

The cumulative effects of such problems can be costly. As of 2001, the estimated cost of the construction of the Solid Fuel Disposition Facility (SFDF) rose from $84 million to $184 million due to problems in obtaining construction and ecological permits from the various state, regional, and local government agencies; reorganization within the Russian government; land estrangement fees demanded by the local government; and relocation of the project from a facility where only renovation was required to one where a complete incinerator needed to be constructed. As this estimate was produced before the latest complications in Votkinsk, the cost of the project is likely to escalate further.[48]

To reduce the risk of cost overruns caused by delays, CTR contractors turned to conducting a wide variety of public relations activities in order to educate the local public on the nature of the projects being pursued in their regions. In some cases, contractors were compelled to deal with issues presented to them by local grass-roots organizations and hold town meetings to explain projects to the local inhabitants.[49] In Votkinsk, the contractors attempted to counteract the environmental resistance to the project by performing educational activities in order to improve public understanding of the nature of the projects being implemented. They also opened offices near local libraries in order to provide further information and held discussions with local politicians to improve their understanding of the projects.[50] Similarly, during work on the elimination of SS-18 ICBM missile silos belonging to the disbanded SRF missile division in Aleysk (Altay Kray) and subsequent site restoration, U.S. contractors attempted to assuage local concerns by presenting detailed explanations of the nature of their work.[51]

---

[46] "Russian Stop-Work Order Averted on Mayak Storage Site," *Post-Soviet Nuclear & Defense Monitor* 5, April 27, 1998, p. 2.

[47] Mikhail Klasson, "Secret trains," *Vremya*, November 23, 2001.

[48] *Cooperative Threat Reduction Annual Report to Congress*, U.S. Department of Defense, 2001, p. III–3.

[49] Interview with U.S. Department of Defense SOAE technical program manager, Washington, D.C., August 2001.

[50] Interview with senior U.S. Department of Defense official, Washington, D.C., August 2001.

[51] Ibid.

Regional Infrastructure Issues

CTR projects implemented on the territory of Russian military bases or contracted to Russian defense enterprises have had to deal with the consequences of chronic lack of support of these facilities by the Russian government. Not only has the number of defense orders placed by the Russian government sharply decreased, but, in many cases, the government has failed to pay for the orders it has placed. Many defense enterprises have thus become heavily indebted to the local utilities and its own work force, and occasionally have faced utility service shutoffs and worker unrest.

The list of examples of affected facilities is long. Power grid failures resulted in electricity shutdowns at Mayak and Novouralsk. Bolshoy Kamen, the site of SSBN elimination activities for the Pacific Fleet, was faced with the threat of electricity cut-offs in November 2000. Local power providers owed more than 81 million rubles to electricity-producer Dalenergo at the time.[52] On September 25, 2002, Bolshoy Kamen was cut off for three hours due to city debts.[53] The Sevmash Shipyard, also involved in SSBN elimination in the Northern Fleet, suffered from similar problems. In September 2000, the Arkhangelsk regional power supply company Arkhenergo reduced hot water supplies to Sevmash due to unpaid debts.[54] As of January 2001, Sevmash owed the local power utility Arkhenergo over 100 million rubles, and, as a result, its provision of heat was reduced by 40 percent during that frigid month. Some of the enterprise's auxiliary facilities suffered reductions in electricity service as well. Arkhenergo experts, however, remained assured that the reduction in electric power supply would not compromise Sevmash's radiation safety procedures.[55] In both cases, the problem was due to the Russian government's outstanding debts to the enterprise. Although the Russian government pledged to pay off the 2 billion ruble debt to the enterprise, as of October 2002 it had not made any payments.[56] Such shutoffs have threatened a wide range of enterprises. The Zlatoust Machine-Building Plant, which is involved in ballistic missile elimination, owes 217 million rubles (the

---

52 "Neskolko gorodov Primorya mogut ostatsya bez elektroenergii iz-za dolgov pered 'Dalenergo'," ITAR-TASS, October 27, 2000; in Integrum Techno, <http://www.integrum.ru>.

53 Radio VBC, as cited in "Primorskiy kray. Na tri chasa ostalsya bez elektrosnabzheniya Bolshoy Kamen," Regions.ru website, September 26, 2002, <http://www.regions.ru/>.

54 "Ya ne to eshche skazal by," *Pravda Severa,* No. 166, September 12, 2000; in Integrum Techno, <http://integrum.ru>.

55 "Sankt-Peterburg: Segodnya s 15.00 'Arkhenergo' vvedet dopolnitelnyye ogranicheniya na postavku teploenergii GP 'Sevmashpredpriyatiye'," RIA RosBiznesKonsalting, January 31, 2001; in Integrum Techno, <http://www.integrum.ru>.

56 "Rossiyskiy tsentr sudostroyeniya okazalsya v slozhnoy situatsii iz-za dolgov voyennykh," Interfax, May 12, 2001.

largest electricity debt owed by any enterprise in the Chelyabinsk Oblast) to Chelyabenergo, the Chelyabinsk energy utility.[57]

By and large, however, the Russian power grid has proved reliable, as long as payments for the service have been made. Nevertheless, projects that have been run by U.S. firms have sought to further reduce this risk by using more than one electrical supply grid for the purpose of redundancy.[58]

The issue of utility services is further complicated when more than one Russian agency participates in project implementation. At the Surovatikha ICBM elimination base, for example, a problem arose with the power supplies. While the site belonged to the MOD, the elimination facility there belonged to the RASA. To prevent complications, the CTR project assumed responsibility for maintaining electricity supplies for the dismantlement activity.[59]

The magnitude of this problem is not constant, but rather varies with the economic fortunes of the Russian Federation. Following the ruble crash of August 1998, some facilities receiving CTR assistance found themselves unable to afford electricity and were, as a result, disconnected from the power grid, putting the security of these installations at risk. This development forced the U.S. government to make some changes to its assistance programs and, for example, emphasize making "quick fix" improvements to physical security systems at facilities with fissile materials before proceeding with complex system upgrades.[60] It is likely, in the event of another sharp economic downturn, that this problem would once again acquire greater urgency.

The chronic nonpayment of wages has led to occasional worker unrest. The Nerpa Shipyard in Murmansk experienced worker hunger strikes in 1998, caused by an eight-month delay in salary payments, caused by the MOD's debt to the shipyard of $12 million.[61] During the same year, Zvezda Shipyard workers in the Far East threatened acts of civil disobedience, went on strike, and briefly blocked the Trans-Siberian Railroad.[62] Severodvinsk's nuclear shipyards, which were owed 1.8 billion rubles by the Russian government, were hit by a number of short strikes

---

[57] RIA-Novosti, November 18, 2001; in "Energy firm threatens to cut power to Russian missile enterprise," FBIS Document CEP20011118000050.

[58] Interview with senior U.S. Department of Defense official, Washington, D.C., August 2001.

[59] Interview with congressional staffer, Washington, D.C., August 2001.

[60] Interview with Rose Gottemoeller, Washington, D.C., September 2001.

[61] *Vesti*, May 18, 1998; in "Russian TV Shows Striking Subyard," FBIS Document UMA-98-139.

[62] Larisa Beloivan and Anatoly Ilyukhov, "The Trade Union Federation of the Maritime Territory Approved the Decision of the Personnel of the Local Enterprises of the Defence Industry to Hold an Action of Civil Disobedience," RIA Novosti online edition, <http://www.ria-novosti.com/ruproducts/hotline/1998/06/02-017.html>, June 2, 1998.

in 1998.[63] The shipyards were forced to provide the workers with foodstuffs as partial payment.[64]

Organized Crime and CTR Projects

Despite the widespread extent of organized crime in Russia, U.S.-funded assistance projects have not experienced major difficulties. In case of the CTR program, U.S.-conducted Audits and Examinations (A&Es) of CTR activities in Russia have determined that the supplied equipment was in nearly all cases being used in accordance with the original U.S. intent, indicating a low level of penetration by local criminal interests. The high degree of project integrity is also attested to by the numerous site visits of CTR program managers and individual contractors (over 400 during FY 2000 alone).[65]

Organized crime poses a more indirect threat to projects by potentially placing in jeopardy the financial well being of enterprises receiving U.S. assistance and even threatening key personnel at such facilities. A number of Russian enterprises participating in CTR projects have had contacts with organized crime in the past. In 1999, to cite just one example, three senior Sevmash officials were sentenced to five years in prison for illegally selling metal products to Sweden.[66]

The DOE Experience

Regional factors were also evident in the implementation of DOE projects, although DOE's experiences with these factors differed from DOD's due to the character of DOE programs, which in turn affected the nature of DOE's working relationship with Minatom.

Although in implementing the MPC&A program DOE also deals with the MOD, Minatom is the most important federal-level Russian entity involved in other major efforts, such as the DOE's Initiatives in Proliferation Prevention (IPP) and Nuclear Cities Initiative (NCI). Many of the DOE-Minatom projects are implemented in the closed nuclear cities around nuclear research and/or production facilities. Due to their sensitive nature, Russian nuclear cities have been since their creation isolated communities supported directly by the central government and its respective ministries. Commerce and the movement of people between closed cities and surrounding oblasts was limited or nonexistent. At the same time, the

---

[63] Vladimir Anufriev, ITAR-TASS, October 7, 1998; in "Severodvinsk Shipyards Stage 1-Day Strike 7 Oct," FBIS-SOV-98-280.

[64] *Russian Public Television First Channel Network*, September 9, 1998; in "Work Suspended at Russian Nuclear Submarine Shipyards," FBIS Document UMA-98-252.

[65] *Cooperative Threat Reduction Annual Report to Congress*, U.S. Department of Defense, 2001.

[66] "Strategicheskiy metal uplyl v Shvetsiyu," *Pravda Severa*, July 15, 2000; in National News Service website, <http://nel.nns.ru>.

cities enjoyed a higher standard of living and services than surrounding areas, and defense establishments around which they were built were considered highly prestigious places of work. The end of the Cold War meant both a relaxation of the restrictions, but also a deterioration of their privileged status, due to the economic crisis. As a result of lower federal funding, interaction with the outside world became not only possible but also a matter of economic necessity. Some nuclear cities have entered into lucrative deals with local commercial partners in order to compensate for the shortfalls in funding. However, in providing financial benefits to the cities, such dealings also encourage the cities to undertake decisions independently of Moscow.[67]

During the 1990s, these entities experienced an increase in local autonomy. The economic autonomy of the closed cities was actually encouraged by the Russian government by the amendment of the 1998 law "On Closed Territorial Administrative Entities" to allow nuclear cities to retain tax revenues, rather than remit them to the federal government. The goal of this amendment was to allow the nuclear cities to reinvest the revenues and spur local manufacture of consumer goods, thus generating profits.[68] However, while the program was partially successful in improving the financial situation of the closed cities, it also led to charges of corruption being leveled at local politicians for allowing outside firms to register in closed cities in order to avoid taxes. In 2001, the tax privileges extended by the 1998 amendment were revoked.[69] This revocation of closed cities' tax privileges was consistent with President Putin's attempts to reverse many of the gains made by the regions during President Yeltsin's rule, to bring regional constitutions in line with federal law, and to eliminate bilateral agreements concluded between the center and the provinces. As with Putin's other attempts to reduce regional autonomy, this one also has not been fully successful.

The experiences and perceptions of DOE in dealing with regional factors have differed from those of DOD for a number of reasons. These include the differing nature of their projects. As mentioned above, DOD projects, as a rule, have concerned themselves with eliminating a set number of delivery vehicles or infrastructure. As a result, DOD projects have more easily measurable success criteria and have therefore not been as open-ended as DOE projects. DOE projects have tended to have farther-reaching and less easily quantifiable goals. The NCI program seeks to create non-military occupations for specialists from the former Soviet nuclear weapons establishment, and the IPP program has attempted to tap into the Russian nuclear laboratories' scientific expertise by establishing non-defense-related business collaboration between Russian nuclear laboratories and U.S. industry. In contrast to most DOD programs (and DOE's MPC&A program), NCI and IPP do not have a well-defined final objective, have open-ended timeframes, and consequently require greater interaction between the

---

[67] Interview with Congressional staffer, Washington, D.C., August 2001.

[68] Elena Sokova, "Russia: Nuclear Cities Initiative," Nuclear Threat Initiative website, <http://www.nti.org/db/nisprofs/russia/forasst/doe/closcity.htm>, April 19, 2002.

[69] Ibid.

implementing agencies of both countries. Moreover, whereas the DOD projects (and DOE's MPC&A program) do not seek to make fundamental changes in the structure of the Ministry of Defense, IPP and NCI could potentially lead to significant alterations in the operations of nuclear closed cities and Minatom itself.

In the case of the MPC&A program, once the amount of money, roles and responsibilities of both governments, and which facilities are to receive MPC&A upgrades are determined at the federal level by a government-to-government agreement, local officials have a relatively high freedom of action regarding implementation. Therefore, the most important factor is the working relationship established between individual U.S. officials and contractors and the specific facilities. In the case of Minatom facilities, the key officials whose cooperation is instrumental are the facility director, the local FSB senior official, and usually the chief of security at the facility. Local authorities have proven considerably more willing and able to attempt to overcome federal-level obstacles to program implementation than in the case of other DOE programs. Contracts for services and equipment are also issued by the U.S. teams directly to their Russian counterparts, without Russian federal involvement. Naval installations, on the other hand, are subject to more centralized control and, consequently, the local officials are correspondingly less powerful.[70]

In the cases of the IPP and NCI programs, DOE has tended to leave decisions concerning the details of resource use, finances, and personnel (areas where regional factors, as seen above, can play a significant role) to the Russian side, while preferring to focus on major policy matters. DOE's chosen route for dealing with regional factors for the NCI program, however, was through Minatom channels. In regard to IPP, Minatom plays no official role but seems to have considerable influence. This has led to a differing perception of regional problems by DOD and DOE officials. Since DOE left regional issues to be dealt with by Minatom itself, its officials did not acquire the same level of awareness of these issues as their counterparts in DOD.

Consequently, DOE officials involved in IPP and NCI projects in Russia have tended to view Minatom as the single most important factor in the implementation of assistance projects and tended to downplay the role of regional factors. However, DOE officials have also acknowledged that relevant personalities can be a more important factor than the institutions involved. This has been reflected in the importance attached to individual personalities, such as the attitudes of institute directors toward cooperation with the United States, and the ability of key local officials to work together.[71] In the cases of IPP and NCI, the key personnel also include local government officials, and in particular the mayors of the closed nuclear cities. Beyond the views of the institute directors, strong ties to the city

[70] Interviews with DOE officials, cited in Sharon K. Weiner, "The U.S. Experience in Implementing Department of Energy Nonproliferation Programs in Russia's Regions," unpublished paper, June 2002.
[71] Ibid.

mayors and administrations have resulted in a "business friendly" attitude in Sarov and Zheleznogorsk.[72]

Although Minatom and other federal agencies are viewed as being the single most important factor in the implementation of assistance programs, it also appears that Minatom's influence varies from one nuclear city to another. The degree of leeway nuclear cities and facilities can exercise also depends on the degree of importance Minatom attaches to a program. The relatively low emphasis Minatom puts on the NCI program, for example, means fewer directives from Moscow on its implementation. The evidence, however, is not clear on whether the distance separating nuclear cities and Moscow is a factor. For example, the relatively more cooperative attitude of authorities in Zheleznogorsk in implementing IPP and NCI projects has been attributed to the city's distance from Moscow.[73] Furthermore, the experience with implementing the MPC&A program at various Minatom facilities scattered across Russia indicates that Minatom is harder pressed to oversee closely the sites that are further away and that distance often benefits the program.[74] However, Snezhinsk's greater distance from Moscow than Sarov's did not result in a more cooperative attitude on the part of Snezhinsk.[75]

It should also be noted that the relative influence of closed city mayors and nuclear facility directors varies depending on the level of federal funding of the facility in question. During the late 1990s, when the Russian government was not able to fund its nuclear facilities adequately, and the closed cities were awarded tax privileges, the relative power of city authorities grew. In 1999 and afterwards, when the Russian government was able to reestablish a sense of financial security at the nuclear facilities, the increased economic power of the nuclear facilities meant their influence in local decisionmaking increased. The level and regularity of federal funding also influence the readiness of nuclear facilities to participate in DOE programs. Nuclear facilities that receive regular federal financing have proven to be less interested in such programs as NCI. Moreover, in many cases, nuclear facility personnel view projects unrelated to their "core mission" (nuclear fuel or warhead work) as less glamorous and less desirable.[76]

In Sarov, the city government and VNIIEF appear to share common interests, although this may be at least in part due to the primacy of VNIIEF over the local government. The government of Sarov has developed defense conversion plans and municipal funding schemes and collaborates with VNIIEF to provide employment opportunities for laid-off workers. VNIIEF's conversion department has also had a positive influence. Another Sarov facility, however, the Avangard Mechanical Plant, appears to be willing to take fewer risks in regards to conversion and has not developed its business skills to the same extent.

---

[72] Ibid.
[73] Ibid.
[74] Ibid.
[75] Interview with Elena Sokova, Monterey, CA, October 2002.
[76] Ibid.

The relationship between Zheleznogorsk and the institute is of a cooperative nature, with leaders of both being interested in economic diversification. Nevertheless, the IPP and NCI projects have had less success here than in other closed cities. This may be caused by Minatom's opposition to transferring technologies from the Mining and Chemical Combine, which was involved in plutonium production, into the open part of the city for development. Another cause may be the incompatibility of high-technology projects with the blue collar nature of the institute's earlier defense work.[77] Others, however, voiced the view that Zheleznogorsk's greater emphasis on applied science actually made it easier to work with.[78]

The situation is less clear in Snezhinsk. The increases in funding for weapons-related work at the All-Russian Scientific Research Institute of Technical Physics (VNIITF) and the failure of its conversion venture Spektr may be the cause of the lower degree of interest in nonproliferation programs in evidence there. However, the less cooperative attitude may have been due to the hostility between the VNIITF director, who favored a continued emphasis on weapons research, and the city's mayor, who is more interested in economic development and defense conversion. Moreover, VNIITF may view the IPP and NCI programs as helping the city become more independent of the institute, which during the Soviet era had more influence than the city's government.

The availability of nuclear weapons-related work is also a factor affecting cooperation with the United States. In the case of the MPC&A program, the facilities that have experienced the slowest progress are the ones that receive revenue from blending down highly enriched uranium (HEU) as part of the U.S.-Russian HEU Deal. Major salary boosts in the past few years at VNIITF and other nuclear facilities have reduced the luster of Western programs.[79]

In addition to cooperation between facility directors and closed city officials, there are instances of cooperation between directors and oblast-level authorities. The good relationship between the director of VNIIEF and the oblast governor exists thanks to the latter's view of VNIIEF as a science center that could revitalize the entire oblast. At one point, oblast authorities were trying to encourage investment through the "Cities of the Volga" campaign. In spite of the potential, the regional relationship is so far limited to one or two business contracts. For example, Nizhniy Novgorod purchased a small number of road repair trucks from VNIIEF that use a patching technique developed through the NCI program.[80]

However, regions will not be able to play a significant role in developing the nuclear cities until there is an economic revival in all of Russia. Nuclear cities still enjoy much better conditions than adjacent oblasts, removing incentives for

---

[77] Ibid.

[78] CNS staff interview with DOE official, March 2000.

[79] Interviews with DOE officials, cited in Sharon K. Weiner, "The U.S. Experience in Implementing Department of Energy Nonproliferation Programs in Russia's Regions," unpublished paper, June 2002. Ibid.

[80] Ibid.

integration. Moreover, there is residual hostility toward nuclear cities for their past privileges.[81]

Conclusion

The experience of U.S. government agencies in implementing nonproliferation assistance projects in Russia has shown the great extent to which regional authorities and other regional actors can influence the course of particular projects. Failure to establish good working relations with the key regional players can result in project delays and cost escalation. In the course of several years of work, U.S. government agencies and contractors participating in the projects have gained immense experience in enlisting the aid of local governments and in coordinating the activities of the various government agencies involved. Although in some cases regional factors hampered project implementation, many project successes in the past were at least partly due to the willingness of regional actors to make decisions in the absence of guidance from the center. Regional Russian actors have been able to lobby for assistance projects, an activity that the U.S. government and U.S. contractors cannot perform. Moreover, experience has shown that the consolidation of central authority has not always led to greater progress in project implementation. The growth of the influence of the intelligence services under President Putin, in particular, has led to greater reluctance by regional governments to exercise freedom of action. Some U.S. officials have expressed concern that the window of opportunity for nonproliferation assistance programs is closing. There are indications that the attitude at the highest levels of the Russian government toward assistance programs has become less favorable. In December 2003, Russia's Security Council Deputy Secretary Oleg Chernov announced that the council was debating whether international assistance for nonproliferation was compatible with Russia's national security interests. Chernov remarked that all too often assistance programs were used as a means to pressure Russia by attaching various political conditions.[82] It is likely that local-level officials hearing such remarks will become less willing to cooperate with U.S. agencies.

As the DOD and DOE experiences over the course of the last decade show, regional factors are in a state of constant flux. As the Russian political system continues to evolve and solidify, the extent and character of regional influences will change with it. Barring a severe economic or political internal crisis, it appears likely that the situation of ill-defined areas of responsibility of local, regional, and central government, and the overall weakness of governments at all levels that prevailed for most of the 1990s will most likely not endure. It is not yet clear whether the Russian political system will evolve in the direction of greater centralized control, as intended by President Putin, or whether it will formally give

---

[81] Ibid.

[82] "Sovbez: Rossiya namerena peresmotret svoi programmy po unichtozheniyu OMU," Grani.ru website, December 3, 2003, <http://www.grani.ru/War/Arms/m.52795.html>.

greater powers to regional and local governments. Should the balance of power continue to swing back in the direction of greater centralization, future nonproliferation assistance efforts in Russia may all follow the DOE pattern, with a stronger central government role in resolving regional issues.

Chapter 11

Nuclear Decentralization in Russia: Lessons Learned and New Directions

Adam N. Stulberg

This book has demonstrated that actors below the federal level do indeed matter for understanding politics and policymaking in Russia's nuclear sector. Although the authors addressed different dimensions—including military and commercial programs, federal and local level issues, and formal legal authority versus actual behavior—their findings highlight two shared conclusions regarding the decentralization of Russian nuclear policy. First, political, economic, and social factors below the federal level play critical roles at both the front and back ends of nuclear policymaking and implementation. A second collective finding is that evidence of decentralization in Russia's nuclear policymaking is not tantamount to a system-wide loss of control over the nuclear sector, despite specific (and sometimes serious) problems in managing material and technology in a number of particular regions and facilities. Thus, U.S. policymakers trying to understand the threats posed by nuclear decentralization should find grounds for real concern, but they should not panic and proceed to the conclusion that the situation has gotten so out of hand that effective policy solutions cannot be crafted to address them. The necessary tools, however, lie not only in Washington and in the other capitals of assistance providers, but also in Moscow and in the regions themselves. In this regard, cooperative approaches may have the best chances for success.

Taken together, the chapters presented here offer considerable evidence of regular and multifaceted involvement of oblast- and local-level political and economic actors in the formulation and implementation of Russia's nuclear policies and related cooperative assistance programs. Since the Soviet collapse, this interface has ranged from threats issued by regional politicians to commandeer strategic nuclear forces; to attempts at flaunting centrally imposed nuclear safety and reform measures for purposes of rent seeking, economic relief, or regional politics; to the formation of regional-industrial lobbies at the federal level forged around mutual interests in expanding nuclear power generation. There also have been numerous acts of personal desperation, born out of frustration with neglect of local conditions by bureaucrats in Moscow. Moreover, these activities have persisted in the face of Russian President Vladimir Putin's concerted campaign to strengthen the vertical chain of command across Russia. But the authors also share the conclusion that the effects of decentralization have not been uniformly disruptive, and in some instances have bolstered transparency and oversight of Russia's nuclear complex. Regional involvement at times has abetted implementation of cooperative assistance programs, removing political and economic barriers at the federal and local levels. In short, while decentralization in Russia is a new and unexpected phenomenon facing Western aid providers, it may manifest itself in different ways: some harmful and others benign. While decentralization challenges the integrity and stability of the nuclear complex, it does not necessarily undermine it, depending on how it is managed. These conclusions suggest that the linkages between decentralization and nuclear security are important themes to explore further in this chapter as well as in future research.

The evidence presented in this book of decentralization in the Russian nuclear complex warrants reassessment of the strategic implications associated with the breakdown in central authority by scholars and policymakers alike. To date, the issues and plethora of analysis tied to fiscal federalism, new forms of local governance, and oscillations in center-periphery relations in Russia have not seeped into discussion or practical thinking about the stability and security of the nuclear sector. While analysts have identified the conditions that confront specific nuclear institutes and touched on civil-military ties at the local level, there has been a dearth of systematic inquiry into how and to what effect regional political offices "matter" for managing Russia's nuclear complex.[1] Policymakers and Western assistance providers tend to dismiss the peculiarities of sub-federal intervention, and to gloss over the role that these actors play in policy reviews of specific programs. There is a proclivity to lump together all forms of subversion as rooted in bureaucratic opportunism, the entrepreneurial skill of specific nuclear facility directors, and poaching by criminal conspirators.[2] Yet, as the chapter authors demonstrate, these assumptions are outdated. Furthermore, given the national

---

[1] Susan L. Clarke and David R. Graham, "How Near to Collapse is Russia Itself?" *Orbis* 39 (Summer 1995), pp. 329–351; Dale R. Herspring, "The Russian Military Faces 'Creeping Disintegration'," *Demokratizatsiya* 7 (Fall 1999); Eva Buzsa, "From Decline to Disintegration: The Russian Military Meets the Millenium," *Demokratizatsiya* 7 (Fall 1999); Valentin Tikhanov, *Russia's Nuclear and Missile Complex: The Human Factor in Proliferation* (Washington, D.C.: Carnegie Endowment for International Peace, 2001); Oleg Bukharin, "The Future of Russia's Plutonium Cities," *International Security* 21 (Spring 1997), pp. 126–158. For earlier accounts of the interface between local politics and nuclear issues in the Russian Far East, see James Clay Moltz, "Russian Nuclear Regionalism: Emerging Local Influence over Far Eastern Facilities," in Judith Thornton and Charles E. Zeigler, eds., *Russia's Far East: A Region at Risk* (Seattle: University of Washington Press, 2002), pp. 247–266; and Cristina Chuen and Tamara Troyakova, "The Complex Politics of Foreign Assistance: Building the Landysh in the Russian Far East," *The Nonproliferation Review* 8 (Summer 2001), pp. 134–149.

[2] U.S. General Accounting Office, *Limited Progress in Improving Nuclear Material Security in Russia and the Newly Independent States*, GAO/RCED/NSIAD-00-82, March 2000; U.S. General Accounting Office, *DOE's Efforts to Assist Weapons Scientists in Russia's Nuclear Cities Face Challenges*, GAO-01-429, May 2001. See also discussion in Scott Parrish and Tamara Robinson, "Efforts to Strengthen Export Controls and Combat Illicit Trafficking and Brain Drain," *The Nonproliferation Review* 7 (Spring 2000), pp. 112–124.

security issues at stake, legacy of monolithic control of the Soviet nuclear sector, and the enduring hierarchical structure of the residual Russian nuclear complex, decentralization of Russia's nuclear complex strikes many as unexpected. Accordingly, regional and local interference is typically dismissed as an *ad hoc* phenomenon, symptomatic of the limits on the Russian state to perform basic functions, including maintenance of stable command and control of nuclear materials and facilities. Once again, the evidence presented in this book challenges this conventional wisdom, demonstrating that decentralization does impact even this most unlikely sector of Russian politics.

Beyond drawing attention to the phenomenon of decentralized policymaking and the imperative for rethinking conceptual and practical assessments of nuclear security in Russia, this book raises two additional issues. First, how do we make sense out of the variations in incidence, dimensions, and effectiveness of sub-federal level participation in nuclear politics across Russia? Are there patterns of influence that cut across different regions and dimensions of governance? Second, what can Russian and Western policymakers do to shore up the effectiveness of administrative control and cooperative assistance in light of the findings presented in the chapters? How might assistance programs benefit from "constructive" regional and local participation without becoming increasingly vulnerable to the deleterious effects of center-periphery politics in Russia?

This concluding chapter offers a preliminary cut at systematically addressing these two lingering issues by explicating the managerial challenges posed by the separation of centralized power and control over Russia's nuclear complex. It reviews the variety of dimensions to oblast- and local-level influence on nuclear policymaking highlighted in the chapters, with specific reference to the different consequences for administrative control and nonproliferation concerns. There is discussion of how weak and vulnerable regional leaderships create conditions ripe for official and unofficial elements to exploit the nuclear sector. Moreover, it demonstrates how strong sub-federal administrative bodies enjoy considerable leeway to subvert federal policies at the local level, while constraining the opportunism of federal agencies and expediting cooperative nonproliferation assistance programs involving foreign countries. Consequently, this analysis gives pause to the pervasive pessimism regarding the link between decentralized power and control problems in Russia's nuclear complex, but also is less sanguine about the prospects for Putin's recentralization reforms at insulating Russia's nuclear policies from the rough and tumble of regional politics.[3]

---

[3] For distinctly pessimistic accounts, see especially Stephan Blank, "Russia as Rogue Proliferator," *Orbis* 44 (Winter 2000), pp. 91–107; and Ariel Cohen, "The Centralization of Russian Rule and its Regional Military Industrial Complexes: The Challenges of Strategic Weapons Proliferation," Chapter presented to The Nonproliferation Policy Education Center for the Workshop on Reassessing U.S.-Russian Strategies Against Strategic Weapons Proliferation, unpublished, June 12, 2001). For cautious optimism, see especially Siegfried S. Hecker, "Thoughts About an Integrated Strategy for Nuclear Cooperation with Russia," *The Nonproliferation Review* 8 (Summer 2001), pp. 1–24.

This chapter proceeds first with a summary of the basic findings presented in earlier sections of the book. The focus is on common themes related to general uncertainties surrounding the ability of Russian policymaking to manage the nuclear complex. The second section reviews common conclusions related to persistent gaps in the distribution of power and responsibility for oversight, specifying three dimensions of regional intrusion into Russian nuclear policymaking. The chapter then assesses the various implications that decentralization has for securing and stabilizing Russia's nuclear sector. The final section offers practical lessons for refining cooperative nonproliferation assistance given these findings.

Uncertainty Amid Hierarchy in the Russian Nuclear Complex

A central finding from the chapters in Part I of the book is that in contrast to Soviet practice, where the state was the sole proprietor of the entire nuclear sector, the Minatom complex is characterized by variable managerial profiles. Ownership of Russia's nuclear facilities, bifurcated by presidential decree in 1993, consists of state unitary facilities and research centers on the one hand, and quasi-private entities on the other. As traced by Orlov and Evstafiev, the formal legal context has devolved decisionmaking and oversight responsibilities to regional and municipal governments. They show, in particular, that by 1995 the federal government was virtually stripped of exclusive control over the nuclear complex, as both "joint" and "independent" responsibilities were legally ceded to regional administrations. At the same time, municipal bodies were empowered to decide issues bearing on the location and construction of nuclear facilities, as well as to generate local environmental impact studies, inform the public of prevailing radiation levels, and adopt measures to protect citizens and their property in the case of a nuclear accident. The effect of these legal changes, the authors conclude, has been to delegate to regional and local politicians considerable but ambiguous jurisdiction over the formulation and implementation of national nuclear policies.

Adding to the confusion, institutional uncertainties confounded longstanding informal mechanisms for managing federal nuclear agencies. Previously, the Soviet leadership maintained formal control through the nuclear sector's priority status at the highest political level, strict and intrusive oversight by the Communist Party and security services, preferential budgetary allocations, and the personal accountability of nuclear institute and enterprise directors. These controls were replaced in the "new Russia" by a combination of benign neglect among the national leadership, budgetary austerity, and a push to make the industry search for new ways to maintain economic viability. Ambiguities associated with "joint jurisdiction" among federal agencies complicated oversight and provided occasion for bureaucratic entrepreneurship. Amid the administrative disorder, for example, Minatom seized opportunities to advance an ambitious strategy for increasing nuclear power generation at the expense of rival government agencies. This included a plan for consolidating the ministry's control over profit-generating civilian nuclear research and defense-related nuclear production.[4] Ben Ouagrham claims that the absence of clearly delineated regulatory responsibility also paved the way for Minatom to expand its jurisdiction at the expense of other ministries and federal regulatory bodies. Overlapping mandates between the Ministry of Defense, Minatom, and the FSB similarly complicated access to otherwise "open" nuclear facilities in the defense and commercial sectors. According to Jasinski, this arrangement exacerbated managerial difficulties by adding multiple layers of bureaucracy for issuing clearances to nuclear facilities and closed cities that are governed by different sets of interests and criteria for "preserving secrecy."

At the same time, Ben Ouagrham argues that the de-institutionalized policymaking landscape complicated vertical control within the Minatom apparatus. On the one hand, the Soviet legacy of hierarchy in the nuclear industry was passed on to Minatom. In particular, the ministry inherited several direct and indirect instruments for coercing and inducing compliance from nuclear facilities across Russia. Formal control mechanisms include delegating the authority to supervise sales of nuclear material from the national stockpile, certifying nuclear operators, setting standards for monitoring and enforcing quality control at nuclear facilities, providing technical approval for nuclear exports, monitoring internal compliance with national export control laws, and approving the (de)classification of nuclear information and travel clearances for top-level managers in the nuclear complex. In addition, Minatom has the authority to supervise the management and corporate governance practices of both state-owned and open joint stock companies in the nuclear sector, as well as to adjust priorities, coordination, and allocations of federal financing and foreign assistance for the nuclear industry. According to the authors, the ministry has capitalized on the residual corporate culture of centralization that pervades the nuclear complex, its own strong representation within interdepartmental government commissions, and regular access to senior echelons of the executive branch to usurp authority to request or deny privileges for subordinate nuclear facilities.

Yet, the confusion associated with Russia's transition has had a deleterious effect on Minatom's hierarchical control. The precipitous drop in state defense orders and protracted federal budgetary shortfalls during the first decade of transition called into question the effectiveness of Minatom's financial mechanisms. Throughout the decade, the ministry faced regular difficulties covering operating expenses and salaries in the plutonium complex, and, at times, received less than 50 percent of the annual federal outlays necessary to support defense activities. Even as federal financing for the sector began to pick up in 2000, Minatom's fiscal control had clearly slipped, as salary payments remained

---

[4] Minatom's strategy calls for: increasing power generation from 15 to 45 percent in Russia by 2030 via the construction of as many as three new reactors funded primarily from revenues gained from exports of nuclear technology and profits from long-term storage of foreign-origin spent nuclear fuel. This strategy provoked reactions from other government agencies including the Foreign Ministry, GAN, and RAO EES (the national electricity company).

delayed and unadjusted for inflation, and unpaid leaves, reduced work schedules, downsized production, and difficult social conditions remained the norm. Accordingly, institute directors acquired strong incentives and *de facto* discretion to look beyond traditional vertical channels for extra-budgetary relief.[5] Minatom's overzealous plan for reducing by half both the number of nuclear weapons design-and-production facilities and the overall size of the nuclear workforce by 2005 only reinforced this impetus for regional autonomy.[6] At the same time, political infighting within the ministry compromised its capacity to guide responses to the social and economic crisis afflicting the nuclear complex. Conflicts over funding priorities for civilian power generation, defense production, and reorganization and control of profit-making activities in the nuclear industry sent mixed signals throughout the bureaucracy. Ben Ouagrham demonstrates that attempts by successive ministers to enlist regional support in their campaign to commercialize the civilian branch of the industry and compete for a larger market share in the domestic energy sector, in effect, emboldened local authorities to subordinate the federally mandated defense conversion agenda to parochial entrepreneurial interests. This combination of budgetary shortfalls and conflicting signals created conditions conducive for regional leaders, environmental groups, and enterprise directors to behave more opportunistically than in the past.

Dimensions of Nuclear Regionalism

The contributors agree that the gaps in federal control over the nuclear complex opened new avenues for regional leaders to assert their political interests directly in the policymaking process. Collectively the case studies demonstrate that regional officials have successfully exercised both *de jure* and *de facto* authority over Russian nuclear politics. In this section, I distill from the chapters the three primary channels of influence used by regional political actors at the federal, fiscal, and regional levels since the Soviet collapse.

The Federal Face

Sokov and Stulberg stress how the formation and evolution of the Federation Council, as stipulated by the 1993 Russian Constitution, provided regional leaders with formal but limited power to shape the country's nuclear legislation. Notwithstanding Putin's reform that ended their *ex officio* membership in the

---

[5] In 1997–1998, Minatom could cover only 20 percent of the actual costs of operating the plutonium complex. Bukharin, "The Future of Russia's Plutonium Cities," p. 132. See also discussion in Igor Khripunov and Maria Katsva, "Russia's Nuclear Industry: The Next Generation," *Bulletin of the Atomic Scientists* (March/April 2002), pp. 51–57.

[6] Igor Khripunov, "MINATOM: Time for Crucial Decisions," *Problems of Communism* 48 (July/August 2000), p. 55; Oleg Bukharin, "Downsizing Russia's Nuclear Warhead Production Infrastructure," *The Nonproliferation Review* 8 (Spring 2001), pp. 125–126.

Council, Russia's regional executive and legislative leaders continued to wield influence in molding the legal context for Minatom's policies. While no longer possessing a direct vote in parliament or parliamentary immunity from federal prosecution, regional leaders gained rights to send and recall full-time, professional representatives to the Federation Council. Moreover, as a political sop and partial compensation for their lost parliamentary status, Putin created the State Council as a consultative body for regional leaders to forge collective platforms on policy issues under consideration by the government. Consequently, regional leaders maintained discretion to determine the nature of control over parliamentary representatives and to instruct delegates to devote substantially more time to specific pieces of federal legislation than they themselves could have in the past. In practice, the significance of the parliamentary reform, in terms of making inter-regional coordination more difficult and susceptible to counter-pressure from the State Duma, has been partially offset by the influence of professional lobbyists who pursue pet regional issues with Minatom. This situation created a new type of Russian official whose effectiveness depends on both establishing a constructive rapport with federal officers and upholding regional interests. According to the authors, the incentive to compromise with Minatom is complemented by the credible threat that regional leaders can instruct representatives to selectively vote down legislation proposed by the executive branch. As demonstrated by the politics behind Russia's legislation to import spent nuclear fuel, regional interests are now taken seriously by government officials at the early stages of drafting and negotiating federal proposals in order to avert a public showdown. Consequently, regional interests play more of a role in shaping the content of federal nuclear policies than is otherwise captured by formal parliamentary voting behavior.[7]

Sokov and Stulberg also argue that regional leaders are able to exploit new, albeit weak, parliamentary authority to forge coalitions with each other and with other powerful interest groups to constrain executive branch policymaking at the national level. The politics of Russia's electricity reform is especially instructive in this regard. Because many local industrial policies have remained at odds with the interests of the national power grid, RAO EES, regional leaders have been impelled to tap their parliamentary authority to side with Minatom in the national debate over electrical power reform. As the authors note, affected regions proved adept at creating a pro-nuclear energy lobby within the Federation Council, as well as at using the State Council as a forum to formulate a rival reform program and to petition President Putin to discipline RAO EES. Without authority to dictate the contours of reform, regional leaders successfully wrangled influence over the substance of federal energy reform by using their national offices to bolster political transparency and raise the political costs to President Putin of arbitrarily neglecting their interests in the process.

Finally, the various case studies reveal that the regional governors can use their position to lobby executive branch officials directly. Sokova considers this

---

[7] Interview with a member of the Russian Federation Council (name withheld by request), December 5, 2001, Washington, D.C.

dimension in her discussion of the regional reaction to the mounting tax burden incurred by the Urals Electrochemical Plant. In this case, the plant, which is responsible for blending down weapons-grade uranium into low-enriched uranium as part of the "Megatons to Megawatts" agreement with the United States, failed to meet tax obligations from 1994-99 primarily because the Ministry of Defense did not pay for state defense orders and Minatom hoarded the foreign assistance previously earmarked for the municipality and facility. In response, Governor Eduard Rossel assumed responsibility for directly petitioning Prime Minister Kasyanov to restructure the federal tax debt of a nuclear facility.

Fiscal "ZATO-ism"

Another avenue for sub-federal intervention is related to public financing of Russia's nuclear enterprises. On the one hand, there is evidence that oblast governments have been marginalized in fiscal transactions pertaining to the nuclear complex. By law, the federal government has exclusive authority to "budget" revenues for the "closed administrative territorial formations" (ZATOs) or "closed cities." These revenues are raised through federal tax transfers (profit, personal income, property, and value-added) and grants that are supplemented by the independent "non-tax" income of the ZATOs. This also holds for the 10 nuclear ZATOs, or nuclear closed cities. Unlike typical Russian cities, the nuclear ZATOs do not interact directly with regional governments on public finance issues, except those regarding collections of road and environmental funds.[8] Sokova argues that in some cases, such as Chelyabinsk Oblast (home to four closed cities, three of which are nuclear: Ozersk, Snezhinsk, Trekhgornyy), fiscal separation is magnified by the fact that federal remittances for the closed cities approximate the entire oblast budget. This dependency on federal transfers and grants noticeably increased following the dramatic devaluation of the ruble in 1998.[9]

Yet, as the chapters by Safranchuk and Sokova demonstrate, regional leaders provided critical supplementary and non-budgetary allocations for the nuclear ZATOs. In 1997, for example, the governor of Nizhniy Novgorod took great interest in promoting transformation at Russia's largest federal nuclear weapons design center in Sarov, creating an oblast-level conversion fund to guide commercial and personnel incentives at the institute. More significantly, Sokova stresses that regional authorities weighed in decisively to influence the

[8] Gregory Brock, "Public Finance in the ZATO Archipelago," *Europe-Asia Studies* 50 (1998), pp. 1065–1081. These findings are updated in Brock, "Public Finance in the Closed Cities of Russia," *The Nonproliferation Review* 11 (Spring 2004).

[9] Gregory Brock, "Public Finance in the ZATO Archipelago," pp. 1067–1068. Federal revenues are derived from profit and income tax, taxes on wage funds, VAT, property taxes, and federal grants. "Non-tax" income for a ZATO includes revenues from the sale of government property, revenues from administrative services, local fines. Federal subsidies to the ZATO precipitously dropped during the August 1998 devaluation crisis. At the same time, non-tax revenues increased relative to federal outlays, and regional administration stepped in to provide temporary relief via revenues generated by road funds.

perpetuation of federal tax shelters in the ZATOs from 1997 to 1999, as well as playing a major role in shaping the subsequent decision to eliminate these "off-shore" privileges.[10] In the case of Lesnoy, such efforts created conditions ripe for commercial entities, such as the energy company, Yukos, to craft regional tax evasion schemes. Furthermore, oblast governments have stepped into the breach by taking it upon themselves to restructure the local debts of Minatom facilities.

This is complemented by evidence that municipal authorities have played an even greater role in keeping the ZATOs financially afloat. Unlike regional governments, city officials retain authority to levy independent taxes, including small education and licensing fees and occasional "wildcat" taxes to generate short-term revenue to support local infrastructure. In some cities, such as Trekhgornyy, the mayor assumed the lead in marketing the city's nuclear conversion projects and updating the local infrastructure to attract commercial investment. His success contributed directly to Trekhgornyy's distinction in 2001 as the only nuclear closed city to become self-sufficient without help from the federal budget or foreign assistance. At the same time, local officials have been authorized to designate line-item expenditures for housing and "other" infrastructure construction within the ZATO budgets. According to one study, the attendant fiscal imbalance between federal revenue–generating authority and municipal discretion over expenditures has perpetuated financial shortfalls for the ZATOs.[11]

Jasinski underscores that the same holds for the defense side of the ledger, as municipal authorities provided critical financial relief—via the "sponsorship" of military units—that circumvented federal channels. The city of Orenburg, for example, extended supplementary credits and assumed the burdens of financing the supporting social infrastructure for the strategic missile unit deployed on its territory. In other cities, such as Kozelsk in Kaluga Oblast, municipal authorities subsidized social and educational expenses for a strategic missile division in exchange for employment opportunities for local civilian constituents. As both Jasinski and Chuen discuss, the parameters for nuclear "sponsorship" were conspicuously extended in several cases. Some sponsoring cities not only provided material support for specific nuclear submarine crews, but stipulated recruitment and personnel policies for respective naval units.

---

[10] In 1992, the Duma approved the law *On Closed Administrative-Territorial Formations* that granted special investment zone status to the ZATO. This status permitted a closed city administration to retain all taxes and revenues collected on its territory, and to grant tax breaks to enterprises to stimulate local investment. By 1997, this law had come under attack for creating unregulated tax havens and for depriving the federal government of significant revenue. In 1998, the Duma authorized the Ministry of Finance to review the privileges granted to the ZATO, and amended the original law by specifying criteria for business registration in the ZATO. The law was officially rescinded in 2000.

[11] Gregory Brock, "The ZATO Archipelago Revisited—Is the Federal Government Loosening Its Grip? A Research Note," *Europe-Asia Studies* 52 (2000), pp. 1349–1360.

Local Motion

The case studies of specific regions reveal a third category of sub-federal involvement in nuclear politics that stems from the impact that regional leaders have on the daily operations of military units and nuclear facilities. In particular, they show that the most conspicuous attempts to influence nuclear facilities occurred when sub-federal officials tried to commandeer nuclear weapons deployed on the territory under their jurisdiction. In March 2000, the Volgograd regional assembly nearly passed legislation that would have extended regional control over Russia's largest missile testing range. This maneuver followed a dramatic incident in 1998 in which Alexander Lebed, the governor of Krasnoyarsk Kray, castigated Moscow for the poor conditions faced by local strategic missile units and hinted at placing the division under his regional jurisdiction. Ultimately, the legislation proposed in Volgograd failed by one vote, and Lebed's threat was issued mainly to pry additional funding from federal coffers and as a political gambit to boost his own presidential pretensions. However, both incidents reflect the extremes to which local leaders have attempted to hijack nuclear policymaking.

Less dramatic, but more intimate forms of influence turn on efforts to co-opt military assets to further regional economic development. Jasinski, in particular, discusses how the governor of Saratov Oblast succeeded at pressuring the Russian Air Force to permit the dual use of the Engels Air Force Base, which is home to a strategic bomber division. By gaining the right to use the base for commercial flights, the governor both improved the local infrastructure and acquired a *de facto* voice in base operations. Similarly, several regional authorities reclaimed deactivated nuclear weapons bases and insisted that eliminated silo sites undergo strict environmental remediation before being turned over to the private sector. On occasion, the author argues, these efforts garnered support from local civic and environmental groups that together strengthened the voice of regional interests in military policymaking.

The case study authors also collectively demonstrate that another low-cost avenue for regional intrusion into nuclear policymaking rests with control over the local infrastructure. Shortfalls and delays in federal subsidies, combined with imbalances between local and federal fiscal responsibilities, rendered military installations and Minatom facilities dependent on regional discretion for access to energy, transportation, public health, military housing, and other local benefits and services. Regional officials seized on this predicament to settle local debts and to collect tribute from vulnerable nuclear facilities. Similarly, regional officials share power with various federal agencies to monitor and enforce safe operations at nuclear facilities via oversight of local emergency response teams, law enforcement, export control, and customs offices. Some regions, for example, created security councils to expedite coordination among federal, oblast, and local branches of law enforcement and nuclear regulatory officials. Sokova argues that others, such as Chelyabinsk and Sverdlovsk Oblasts, assumed direct responsibility for providing supplemental financial and technical support to local internal security troops, as well as housing and social benefits for military personnel, in an effort to upgrade safety and security at local nuclear facilities. Moreover, regional assemblies and municipal courts offer venues for local interest groups and environmental lobbies to petition Minatom's policies, as well as for local military commanders and nuclear institute directors to seek legal remedies for shortfalls in federal obligations. Authority to approve oblast or local referenda and to excite popular debate over nuclear and environmental issues also affords regional leaders with conspicuous, *de facto* influence over the implementation of federal nuclear policies.

Strategic Implications

Recognizing the multidimensional regional intrusion, this book raises several questions: What are the consequences of gaps in hierarchical control of the Russian nuclear complex? Does regional input into the policymaking process help or hinder nuclear safety and security in Russia? Does it advance or confound Minatom's parochial bureaucratic interests and the success of cooperative assistance programs?

The administrative confusion at the federal level in Russia suggests that the hierarchical relationships, both between President Putin and Minatom and between the executive leadership and nuclear facilities within the Minatom structure, are marred by gaps between power and control. Both the president and the leadership of Minatom, for example, enjoy formal and informal authority to oversee the activities of respective subordinated offices. At the executive level, the Russian president is interested in promoting the nuclear sector for the country's international prestige and development, while maintaining checks and balances over the sector to ensure stable control and consistency among national security and industrial policies. Similarly, the leadership of Minatom generally seeks to redress the inherited problems confronting the sector and to increase the ministry's political autonomy and commercial stature, both at home and abroad. Such parochial interests include consolidating and converting nuclear assets in the defense sector, as well as expanding nuclear power generation, technology exports, and earnings from the storage and reprocessing of foreign spent nuclear fuel.[12] Alternatively, sub-units within the Minatom structure and respective nuclear facilities spread across Russia are formally obligated to comply with directives for carrying out nuclear operations in return for financial and political remuneration. The rub, however, is that these agencies enjoy advantages of information and expertise that are critical for conducting complex technical policies in the geographically dispersed nuclear complex. Because the executive leadership in the government and atop the Minatom complex cannot perfectly and readily monitor the behavior of the implementing agencies and facilities, they must expend scarce resources inducing and enforcing compliance. Thus, the challenge confronting both

[12] Bukharin, "Downsizing Russia's Nuclear Warhead Production Infrastructure," pp. 116–130; Khripunov, "MINATOM: Time for Crucial Decisions," pp. 51–54.

the Russian president and the leadership of Minatom is how to mitigate these asymmetries to bolster accountability across the dispersed nuclear complex in a manner consistent with respective national and bureaucratic objectives.[13]

In the context of Russian nuclear regionalism, the policy consequences of these information asymmetries depend not only on the stability and consolidation of regional interests, but also on the level (federal or regional) on which they are asserted. Regional leaders who have consolidated political authority and possess clear policy interests are in a good position to subvert or complicate the implementation of Minatom's policies within their regions. This situation can either improve or mar the effectiveness of nuclear stability and cooperative assistance, depending on the specific interests and resources wielded by respective regional authorities. At the federal level, however, the activism of entrenched regional leaders can curtail Minatom's bureaucratic opportunism and contribute to greater transparency and oversight of the national policymaking process. Acting with independent interests and authority, regional assertiveness here can blow the whistle on ministerial shirking. Alternatively, an embattled or vulnerable regional leadership is more apt to be exploited by Minatom to advance favored policies in the region or to contribute to a general loss of control over nuclear assets in the affected territory. Similarly, regional weakness at the federal level loosens parliamentary checks and balances on Minatom's federal activities.

Strong sub-federal authorities that enjoy independent political and economic bases of local power can affect Russian nuclear policymaking, at both the regional and federal levels. Regional leaders can purposefully undermine or expedite implementation of federal nuclear policies, depending on their particular policy concerns and resource endowments. On the one hand, they can practice "soft subversion," exploiting control over the local infrastructure to extract side payments from Minatom or to co-opt the use of specific facilities, with adverse consequences for nuclear safety.[14] Jasinski, for example, argues that this tactic was employed by the independent-minded governors of Ivanovo Oblast and Primorskiy Kray, who took it upon themselves to unplug strategic military units (including an early-warning radar in the case of the latter) from the local energy grid in an effort to unlock delivery of federal subsidies. He also shows that city officials in

---

[13] This is akin to a principal-agent relationship, whereby information asymmetries reinforce gaps between power and control and create divergent incentives among super- and subordinate levels within an administrative structure. See discussion in D. Roderick Kiewiet and Mathew D. McCubbins, *The Logic of Delegation: Congressional Parties and the Appropriations Process* (Chicago: The University of Chicago Press, 1991); and John W. Pratt and Richard Zeckhauser, "Principals and Agents: An Overview," in Ibid., eds., *Principals and Agents: The Structure of Business* (Boston: Harvard Business School Press, 2001), pp. 1–35. For an application to the Russian context, see especially discussion in Alexander Sokolowski, "Bankrupt Government: Intra-Executive Relations and the Politics of Budgetary Irresponsibility in El'tsin's Russia," *Europe-Asia Studies* 53 (June 2001), pp. 547–549.

[14] The terms "soft subversion" is adapted from Hyde, "Putin's Federal Reforms and their Implications for Presidential Power in Russia," p. 737.

Votkinsk took the issue further by both sponsoring a referendum to build a solid rocket fuel incineration plant on their territory and preventing access to the federal facility, until the city was fully compensated for the loss of real estate incurred as a result. While the referendum was subsequently invalidated by the regional Supreme Court, the president of Udmurtiya continued to block construction of the plant until Moscow agreed to provide additional state defense orders to the republic.

Similar bargaining tactics have obstructed commercial nuclear activities as well. For example, Sokova reveals that on several occasions in 2001, the local energy provider in Chelyabinsk significantly reduced the power supply to major defense-related nuclear facilities in the region, including the warhead assembly/disassembly facility in Trekhgornyy, the nuclear warhead design center in Snezinsk, and the Mayak storage facility in Ozersk. The governor of the same province, emboldened by an upturn in the region's economic outlook in 2001, exaggerated the risks of an accidental overflow of radioactive waste along the Techa River drainage basin to pressure Minatom to resume construction of a nuclear power plant in Ozersk. The governor evinced little regard for local environmental concerns, proposing instead that the polluted water be used to cool the plant.[15] Similarly, Chuen argues that the popular governor of the Khabarovsk region repeatedly confounded Minatom by taking a strong stand against the construction of nuclear power plants on his territory. Neither dependent on Minatom for energy supplies nor beholden to the Ministry of Defense for dismantlement assistance, the Khabarovsk governor consistently charted an independent course to maximize local economic gains on nuclear issues in his region. Local politics surrounding the construction of the Japanese-financed floating liquid radioactive waste (LRW) treatment plant in Primorskiy Kray also falls into this category of regional obstruction.[16]

On other occasions, strong regional leaders have played critical roles expediting implementation of conversion and cooperative assistance at local nuclear facilities. Kovchegin, for example, demonstrates that Tomsk officials, having tied the region's welfare to development of the hydrocarbon sector, have assumed an active role in promoting diversification of the large Siberian Chemical Combine located in the closed nuclear city of Seversk. The governor reached a *modus vivendi* with local nuclear directors, whereby his office promised to cultivate favorable socio-economic conditions for the combine and to lobby on behalf of Minatom's strategy for increasing nuclear power generation. In return, the local nuclear industry committed to remedying short-term energy deficits and contributing to the economic resurgence of the region. The governor also championed the Tomsk Regional Initiative that called for direct cooperation with U.S. government agencies to attract foreign investment to the region. According to

[15] See also Charles Zeigler and Henry B. Lyon, "The Politics of Nuclear Waste in Russia," *Problems of Post-Communism* 49 (July/August 2002), pp. 39–40.

[16] Cristina Chuen and Tamara Troyakova, "The Complex Politics of Foreign Assistance: Building the Landysh in the Russian Far East."

Sokova, other regional leaders exploited their autonomy and resources to facilitate local conversion initiatives. Leaders in Sverdlovsk, for example, held Minatom accountable for meeting international safety standards in constructing a new warhead storage facility in the closed city of Lesnoy. Furthermore, Jasinski and Thornton note that on numerous occasions regional and local administrators, eager to woo foreign assistance and to demonstrate autonomy from Moscow, have taken the initiative to relieve local bottlenecks and "bend the rules" to expedite the implementation of specific projects funded by the Cooperative Threat Reduction (CTR) program.

Strong regional leaders also can use their parliamentary authority to constrain Minatom's bureaucratic entrepreneurship consistent with their own interests at the federal level. This tactic holds mixed consequences for the stability of Russia's nuclear industry and the prospects for cooperative assistance. On issues in which regional interests conflict with national security objectives and Minatom's commercial interests, regional interventions can subvert cooperative nonproliferation assistance. In 2000, for example, regional leaders exercised authority in the Federation Council to express reservations about the U.S.-Russian "Megatons to Megawatts" deal, drafting a resolution that revised the agreement and jeopardized a major funding source for the ministry.[17]

Conversely, independent regional leaders have capitalized on their federal powers to improve the effectiveness of national nuclear policies. According to Chuen, the maverick former governor of Primorskiy Kray, Evgeniy Nazdratenko, collaborated with officials in Bolshoi Kamen and representatives of the Zvezda Shipyard to lobby Moscow in support of the CTR submarine dismantlement program. Similarly, as noted by Sokova, the governor and Duma representatives of Chelyabinsk appealed directly to President Boris Yeltsin and the Federation Council in 1999 to pressure Minatom to fulfill financial obligations to the three closed nuclear cities in the region. In 2001, the Chelyabinsk governor again used his clout in Moscow to draw attention to the fiscal laxity exhibited by Minatom and the Ministry of Defense and to negotiate the rescheduling of the 1994-99 federal tax debt for the Urals Electrochemical Plant.

The political battle over Minatom's proposed legislative amendments permitting Russia to store and reprocess foreign spent nuclear fuel illustrates the role of regional offices in promoting accountability and transparency at the federal level. Initially, Minatom seized upon the prospects for earning $21 billion over 10 years in this potential global market by proposing amendments to the 1992 federal environmental protection law that prohibited the import of radioactive waste.[18] Under the original proposal, Minatom was designated exclusive responsibility both to conclude specific contracts and to allocate revenues for the development of the nuclear industry and environmental remediation. This proposal encountered strong

---

[17] Igor Khripunov, "Minatom: Time for Crucial Decisions," p. 52.

[18] For Minatom's early proposal, see *Technical-Economic Basis for the Law of the Russian Federation: On the Proposed Amendment to Article 50 of the Law of the RSFSR, 1999*, <http://www.bellona.no>.

resistance inside Russia, with regional leaders across the country threatening to vote down the legislation in the Federation Council and polls suggesting that more than 90 percent of the population did not support the amendments. As Sokov and Stulberg argue, in an effort to avert a public showdown, Minatom conspicuously offered policy inducements to regions that stood to be directly affected by the program. These concessions, combined with Minatom's extensive lobbying campaigns to trumpet the regional commercial benefits of the program, succeeded in co-opting support from several critical regions, such as Chelyabinsk and Krasnoyarsk. The final consideration of the proposal in the Federation Council thus became a conspicuous "non-event," as the legislation was passed directly to the president for approval without an extensive review.[19] As a concession to regional and popular agitation over the legislative amendments, President Putin personally intervened to create an interagency oversight body, headed by a Nobel laureate, with executive authority to review specific contracts and supervise the allocation of revenues for environmental cleanup. The authors argue that the attendant costs of policy success for Minatom were significant, as the ministry was formally obligated to share profits and to allow greater transparency in crafting and administering specific deals involving the import of spent nuclear fuel. This transparency strengthened the legal grounds for regional authorities and local environmental lobbies to sound the alarm in the future regarding any questionable activities by Minatom in this field, which can increase the ministry's public accountability for Russia's future foreign spent fuel transactions.[20]

On the other hand, politically divided and economically strapped regional leaderships also impact Russian nuclear policymaking. Here, however, the influence is primarily by omission, as weak governing capacities foster local conditions ripe for corruption, and graft to penetrate nuclear activities. From the perspective of nuclear safety, greater local influence over facilities is often negative, particularly when the root causes are political weakness and economic

---

[19] Rather than hold a formal vote on all three amendments, the Federation Council opted to debate only the measure that stipulates how proceeds from the spent fuel import business are to be used for environmental restoration in the regions. All three amendments were passed directly to Putin for signature.

[20] The first two deals that involved the import of spent nuclear fuel from Bulgaria and Hungary that followed the passage of the amended legislation were contested for technical violations of the law in a special Duma hearing and Russia's constitutional Court, respectively. In the Bulgarian case, Minatom had to justify the specific financial and technical arrangements of the deal. In the second episode, the Constitutional Court invalidated the government resolution that allowed the storage of nuclear waste from the Hungarian facility. See especially Yekaterina Pichugina, "In Secret From Another World: We Are Secretly Being Made Into Nuclear Goat-Milkers," *Moskovskiy komsomolets*, October 19, 2001, FBIS Document CEP20011018000242; Svetlana Babayeva, Serge Leskov, and Svetlana Popova, "The Era of Kozloduj," *Izvestiya*, January 22, 2002, FBIS Document CEP20020122000602; "Duma Deputies Make Inquiry About Illegal Import of Foreign Spent Nuclear Fuel," Interfax, December 3, 2001; and "Court Scraps Resolution on Storing Spent Nuclear Fuel From Hungary," Interfax, February 26, 2002.

distress (often prompted by fiscal and administrative abandonment by the center). The net result in these cases is poorly informed and occasionally dangerous behavior by local decisionmakers that flaunts centrally prescribed nuclear safety regulations for the purposes of personal profit, increased energy production, or other rent-seeking purposes. This situation is compounded by the underdeveloped state of democratic institutions in these regions, where well-informed, non-governmental organizations (NGOs) are virtually non-existent and where the media is often weak, poorly funded, and under the control of local enterprises or political machines. These conditions impede information gathering on nuclear matters, thus promoting skewed decisions that serve the parochial interests of powerful regional actors instead of national concerns for ensuring stability within the nuclear complex.

Problems associated with weak regional leadership have been especially conspicuous in Russia's Far Eastern Federal District, given the relative economic hardships and prominence of criminal networks one the one hand, and the high concentration of nuclear materials, on the other.[21] Chuen argues that the situation in Kamchatka clearly illustrates this thesis, as nuclear issues have been at the center of several hotly contested regional political races, and the region's economic woes have given rise to systematic criminal activities involving theft and diversion of nuclear materials. Different political leaders—including the State Duma representative, members of the Kamchatka Oblast Duma, and the Communist Party governor—staked out contending positions regarding the safety and utility of increasing the region's reliance on nuclear power. While the governor tended to side with Minatom's attempts to increase the profile of nuclear power in the region, his political rival in the State Duma stymied progress on other fronts. The latter, in particular, stoked public apprehension by condemning U.S. assistance for submarine defueling and dismantlement in the region and protesting the import of spent nuclear fuel. Amid the political turmoil, Minatom, which assumed formal responsibility for defueling nuclear submarines in 1998, adopted secretive practices and fell behind in financial support for managing the cleanup of radioactive waste in the region, allowing local politicians to bear the brunt of public criticism. At the same time, the criminal situation in Kamchatka became so serious that a local naval commander equated civilian smuggling of military-related materials to an actual "regional conflict," given the possible adverse effects on military readiness. In January 2000, for example, two sailors bribed their way onto the Rybachiy Naval Base and succeeded in removing radioactive calibrating

---

[21] See especially Moltz, "Russian Nuclear Regionalism: Emerging Local Influence over Far Eastern Facilities," pp. 256–257. Primorskiy Kray, at least under former Governor Nazdratenko, is a potential outlier for this analysis. Under Nazdratenko, the region exhibited both strong/independent executive leadership and an intimate fusion of public and private corruption and criminal behavior. For a discussion of the unique features of the region, see especially Peter Kirkow, "Regional Warlordism in Russia: The Case of Primorskiy Kray," *Europe-Asia Studies* 47 (September 1995), pp. 923–947.

plates from a decommissioned nuclear submarine.[22] A subsequent February 2001 trial of 11 indicted sailors also revealed that platinum catalysts had been systematically stolen from most of the decommissioned nuclear submarines at the base. As Chuen reports, local mafia organs typically disseminate flyers to naval recruits specifying black market prices and instructions for stealing precious metals from the Rybachiy base.

Similarly, weak and nuclear energy dependent regions tend to be enlisted to support Minatom's campaigns to take on other federal bureaucracies. While many regional and local governments acquired authority to regulate polluters in their territories, business pressures, corruption, and close ties to Minatom have not augured well for definitive stands in support of independent regulatory authority or championing local environmental concerns.[23] Leaders from some of Russia's most troubled regions have generally endorsed Minatom's long-term strategy for increasing nuclear power generation, notwithstanding local suspicions about the ministry's managerial practices and protests by grassroots environmental groups. Impoverished regions have tended to be even less concerned with nuclear safety than central authorities, particularly when nuclear power required for the local energy grid is involved. A group of nuclear energy dependent regional leaders, for example, went so far as to form the Union of Territories and Enterprises of the Nuclear Energy Complex specifically for lobbying the federal government to expand construction of nuclear power stations and to endorse Minatom's efforts to import spent nuclear fuel.[24]

Conclusions and Recommendations

The above analysis demonstrates not only that sub-federal political actors do indeed matter in the formulation and implementation of Russia's nuclear policies, but that both the direction and scope of their influence vary systematically across regions and levels of policymaking. Different types of regions exercise different forms of influence at the federal and regional levels, with different consequences for nuclear stability and nonproliferation. Those regions that possess available resources and strong leadership can significantly impede or expedite local implementation of federal nuclear policies, depending on respective instrumental concerns. Similarly, they can serve as important "whistleblowers" to shore up monitoring of federal offices and ministries, thus improving national control over the nuclear sector. Conversely, weak and economically strapped regions tend to

---

[22] The sailors nearly succeeded at breaking into the reactor and raising the control rods. Moltz, "Russian Nuclear Regionalism: Emerging Local Influence over Far Eastern Facilities," pp. 256–257.

[23] Charles E. Ziegler and Henry B. Lyon, "The Politics of Nuclear Waste in Russia," p. 35.

[24] Nikolai Sokov, "The Reality and Myths of Nuclear Regionalism in Russia," *Program on New Approaches to Russian Security*, Policy Memo No. 133 (2000); <http://www.fas.harvard.edu/~ponars/POLICY%20MEMOS/Sokov133.html>.

exacerbate problems of "nuclear opportunism," contributing to the general "loss of control" over the nuclear complex. These regions are especially susceptible to corruption and criminal practices that increase risks of theft, diversion, and proliferation of fissile material from local facilities. Politicians from weak and dependent regions also tend not to exercise respective federal authority with the aim of checking or challenging the bureaucratic strategies championed by Minatom or the Ministry of Defense.

The effects of decentralization on specific issues related to nuclear stability and cooperative assistance are also mixed. On the one hand, a central finding of this book is that regional activism is correlated neither with democratic oversight nor improved safety concerns in the nuclear sector. Strong and autonomous regional leaders typically weigh in on nuclear issues to further parochial economic or political objectives, with little regard for environmental safety or stability of the complex. Local public opinion and grassroots environmental movements—much less desperate military personnel—tend to be ignored, unless regional politicians can co-opt respective issues for political or commercial profit. Similarly, regions that have experienced the fewest problems in nuclear control have tended to be those where Minatom or the relatively well-funded Strategic Rocket Forces has been stronger and where specific nuclear facilities have remained relatively insulated from intra-regional politics and the local economy. On the other hand, independent-minded regional leaders have tended to promote conversion and downsizing of the nuclear complex. Motivated primarily by material interest, strong regional leaderships have prodded both Minatom and local facilities to expedite commercial transformation and have served as creative allies in bolstering the effectiveness of cooperative assistance. These twin pressures suggest that decentralization by itself does not offer a panacea for averting the difficult commercial, security, and environmental trade-offs attendant to reform of Russia's civilian and military nuclear complex.

What do these conclusions mean for the West and how might they influence the future of cooperative nonproliferation assistance to Russia? First, they demonstrate the potential limits of issue linkage as a diplomatic tactic for inducing or pressuring compliance with nonproliferation norms. Those instances in the nuclear complex where there is a separation of power and accountability create alternative incentives among poorly coordinated federal- and regional-level political actors. As a result, rewarding or punishing a federal agency for the behavior of a regional or local actor that is not practically under its control is potentially problematic. Consequently, Western partners should take care to craft assistance programs so that specific regional authorities and firms have a stake in the effort, and that all participating Russian recipients shoulder both the costs and benefits of cooperation. In this regard, assistance providers should build on the spirit of the Nuclear Cities Initiative by working with Russian federal and regional authorities to fund projects directly related to retraining personnel and transforming the local nuclear infrastructure.[25]

Second, Western assistance providers should understand the specific political, economic, environmental, and social dynamics in targeted regions and develop an appreciation for the relative strength and instrumental interests of respective regional leaders. Strong and independent-minded sub-federal level officials should be included (together with federal officials and facility leaders) in negotiations and given discrete responsibilities for supervising specific projects. Excluding such regional concerns, inadvertently or otherwise, risks unnecessary project delays and costs overruns. In addition, neglect of regional actors by Western assistance providers can lead to missing opportunities to enlist important allies in removing local obstacles and containing bureaucratic opportunism at the federal level. Many regions of Russia also would benefit from a focus on nonproliferation technical assistance, information, and education to improve the responsiveness of local politicians to citizens' concerns about nuclear safety and security. Alternatively, in weak and dependent regions, Western partners should work closely with Minatom and other federal offices to redress priority concerns related to the physical protection and security of designated facilities. Specific concerns in these territories should be impressed upon Russian central authorities, and the relevant assistance should be directed at strengthening federal oversight and national export controls.

Finally, in light of the challenges posed by the decentralization of administrative control and the multifaceted role of regional leaders, Western providers should jettison narrow accounting standards for measuring the success or failure of cooperative nonproliferation assistance programs. As discussed above, even the most efficiently designed programs can be subverted from either above or below in Russia, despite the best intentions of program administrators and final recipients or strategic imperatives. Moreover, because these assistance programs generate intangible benefits—such as broadening political support among autonomous policymaking circles across Russia and shoring up gaps in coordination and transparency in the nuclear sector—their significance cannot be adequately captured on an accountant's balance sheet. By appreciating that cooperative assistance programs serve as more than vehicles for dismantling the former Soviet arsenal, and by understanding how they can play critical roles in facilitating the institutionalization of effective oversight of the Russian nuclear

---

[25] The Nuclear Cities Initiative (NCI) was established in 1998 as an equal partnership agreement between the U.S and Russia and has been administered by the U.S. Department of Energy. It was intended to "create a framework for cooperation in facilitating civilian production that will provide new jobs for workers displaced from enterprises of the nuclear complex in the 'nuclear cities' controlled by the Ministry of Atomic Energy." Programmatic objectives have been directed at re-profiling scientific activities and downsizing the nuclear weapons footprint at the designated closed cities. *Agreement between the Government of the United States of America and the Government of the Russian Federation on the Nuclear Cities Initiative*, September 22, 1998.

complex, we can fully assess the effectiveness of specific programs. Failure to do so not only risks alienating potential allies in Russia, but could unnecessarily increase the likelihood that the decentralization under way in Russia will degrade control of the nuclear complex.

Index